Girl Decoded

"Artificial intelligence is advancing exponentially. Creating a hopeful, compelling, and abundant future for humanity depends on the work outlined by entrepreneur Rana el Kaliouby in her must-read book, *Girl Decoded*. It is critical that we embrace a future in which technology enhances our most human qualities, specifically empathy, caring, and emotional intelligence."

—PETER DIAMANDIS, founder and chairman of the XPRIZE Foundation and coauthor of *The Future Is Faster Than You Think*

"Emotional intelligence in AI is every bit as critical for machines as it is for humans. No one has done more to create that capability than Rana el Kaliouby. This book tells the remarkable story of her pioneering journey, breaking barriers all along the way. A must-read!"

—ERIK BRYNJOLFSSON, professor at MIT Sloan School of Management and coauthor of *The Second Machine Age* and *Machine, Platform, Crowd*

"Written with kindness, vulnerability, and grace, *Girl Decoded* reveals the tour de force that is Rana el Kaliouby. Her must-read memoir spurs technologists to follow their conscience and emboldens women all over the globe to fight for their dreams."

—DR. KATE DARLING, leading expert in robot ethics and research specialist at the MIT Media Lab

"This lucid and captivating book by a renowned pioneer of Emotion AI tackles one of the most pressing issues of our time: How can we ensure a future where this technology empowers rather than surveils and manipulates us?"

—MAX TEGMARK, professor of physics at Massachusetts Institute of Technology and author of *Life 3.0*

"Rana el Kaliouby is one of the most brilliant and inspiring people I've ever met, but I wasn't prepared for how this book would move me. The reader joins Rana on an uplifting journey from nice, shy Egyptian girl to scientist, entrepreneur, and visionary American leader. We also witness the birth of an idea—that machines can be trained to recognize and respond to human emotion. Rana expresses her vision with clarity and hope, a̶ intelligent mac̶

nd *Inc.*

ABOUT THE AUTHOR

RANA EL KALIOUBY, PhD is a pioneer in artificial emotional intelligence (Emotion AI), as well as the cofounder and CEO of Affectiva, the acclaimed startup spun off from the MIT Media Lab. She grew up in Cairo, Egypt and holds undergraduate and masters degrees in Computer Science from the American University in Cairo, a PhD from the Computer Laboratory at the University of Cambridge, and a Post Doctorate from the Massachusetts Institute of Technology.

Girl
Decoded

*My Quest to Make Technology
Emotionally Intelligent —
and Change the Way We
Interact Forever*

RANA EL KALIOUBY

with CAROL COLMAN

BUSINESS

PENGUIN BUSINESS

UK | USA | Canada | Ireland | Australia
India | New Zealand | South Africa

Penguin Business is part of the Penguin Random House group of companies
whose addresses can be found at global.penguinrandomhouse.com.

First published in the United States of America by Currency, an imprint of Random
House, a division of Penguin Random House LLC, New York 2020
First published in Great Britain by Penguin Business 2020
This edition published 2021

001

Title and part opener pages image by Getty/rzarek

Printed and bound in Italy by Grafica Veneta S.p.A.

The authorized representative in the EEA is Penguin Random House Ireland,
Morrison Chambers, 32 Nassau Street, Dublin D02 YH68

A CIP catalogue record for this book is available from the British Library

ISBN: 978–0–241–45152–6

Follow us on LinkedIn: https://www.linkedin.com/company/penguin-connect/

www.greenpenguin.co.uk

MIX
Paper from
responsible sources
FSC® C018179

Penguin Random House is committed to a
sustainable future for our business, our readers
and our planet. This book is made from Forest
Stewardship Council® certified paper.

*For my mom for teaching me that it's okay
to embrace and celebrate your emotions*

Contents

Girl
Decoded

Introduction

Emotion Blind

A view of human nature that ignores the power of emotions is sadly shortsighted.

—DANIEL GOLEMAN, PHD, author,

   Emotional Intelligence: Why It Can Matter More Than IQ

In the summer of 2017, Jamel Dunn, a thirty-one-year-old disabled man from Central Florida, realized that he had waded too far out into a pond. He shouted to a group of teenagers hanging out on shore for help, but they were unresponsive; they refused to come to his aid. As the father of two flailed in the water and pleaded for someone to help him, the teenagers actually jeered and taunted him, calling him a "cripple" and yelling, "You're gonna die." They did not race into the water to try to save him. Nor did they use their cellphones to call 911. They *did* use their cellphones, however, to record the event. When, after a few minutes, Dunn disappeared beneath the surface, one of them observed, "He just died," and they all laughed.

How do we know all this? The teens not only recorded the incident on their smartphones but also posted the video online, treating the real-life drama that had unfolded before them as a bit of edgy shock video. Otherwise, this summer tragedy of 2017 in Cocoa, Florida, a small town near Orlando, would likely have passed unnoticed. It certainly would never have been picked up by media world-wide. And I would never have heard about it in the Boston suburb where I live. Dunn's sister learned of the video and notified the police, who brought the teenagers into the station for questioning. A police officer told CNN that the teens showed no remorse; in fact, they showed very little emotion at all. "I've been doing this a long time, probably twenty years or more ... I was horrified. My jaw dropped," she said.

Ultimately, the teens weren't charged with any crime: They were not liable under Florida law to provide emergency assistance or even to report the drowning. Nevertheless, their callousness, their casual cruelty, was horrifying. But that alone doesn't explain why this particular incident went viral. I believe that the lack of basic humanity in these teens hit a deep societal nerve, exposing an ugly truth about the world we now live in. Every day, we encounter people who display a similarly shocking lack of empathy, not to mention basic civility.

It is commonplace on social media and in politics, entertainment, and popular culture to see callous, hateful language and actions that even a couple of years ago would have been considered shocking, disgraceful, and disqualifying. As a newly minted American citizen born in Egypt and a Muslim woman who immigrated to the United States with her two children at a time when political leaders were calling for Muslim bans and border walls to keep immigrants out, I am particularly aware of the insensitive, at times vicious, voices in the cyber world. But, in truth, everyone is fair game.

It is not off-limits to troll survivors of gun violence, like the students of Parkland, who advocate for saner gun laws; to shame victims of sexual abuse; to post racist, anti-Semitic, sexist, homophobic, anti-immigrant rants; or to ridicule people whose only sin is that they disagree with you. This is happening in our communities, work-places, and even on our college campuses. Today, such behaviors are dismissed with a shrug. They can even get you tens of millions of followers and a prime-time spot on a cable TV network—or send you to the White House.

What happened in Florida is representative of a problem endemic to our society. Some social scientists call it an "empathy crisis." It is the inability to put yourself in another's shoes and feel compassion, sympathy, and kinship for another human being. This stunning lack of concern for our fellow citizens permeates and festers in the cyber world, especially on social media, and is spilling over into the real world.

We, as a society, are in increasingly dangerous territory—we are at risk of undermining the very traits that make us human in the first place.

More than two decades ago, journalist Daniel Goleman wrote about the importance of empathy in his bestselling book *Emotional Intelligence*. When he argued that genuine intelligence is a mixture of IQ and what we've come to call EQ, or emotional intelligence, he changed our thinking about what makes someone truly intelligent. EQ is the ability to understand and control our emotions and read and respond appropriately to the emotional states of others. EQ, rather than IQ, is the determining factor predicting success in business, personal relationships, and even our health outcomes.

Obviously, you can't experience emotional intelligence without feeling, without emotion. But when we are in cyberspace, "feelings" don't come into play because our computers can't see or sense them: When we enter the virtual world, we leave our EQ behind.

Inadvertently, we have plunged ourselves headfirst into a world that neither recognizes emotion nor allows us to express emotion to one another, a world that short-circuits an essential dimension of human intelligence. And today, we are suffering the consequences of our emotion-blind interactions.

Computers are "smart" in that they were designed with an abundance of cognitive intelligence, or IQ. But they are totally lacking in EQ. Traditional computers are emotion blind: They don't recognize or respond to emotion at all. Around twenty years ago, a handful of computer scientists—I was one of them—recognized that as computers became more deeply embedded in our lives, we would need them to have more than computational smarts; we'd need them to have people smarts. Without this, we run the risk that our dependence on our "smart" technology will siphon off the very intelligence and capabilities that distinguish human beings from our machines. If we continue on this path of emotion blind technology, we run the risk of losing our social skills in the real world. We will forget how to be compassionate and empathetic to one another.

I am cofounder and CEO of a Boston-based artificial intelligence company, a pioneer in Emotion AI, a branch of computer science dedicated to bringing emotional intelligence to the digital world. Artificial intelligence, or AI, is the science of training computers to think and reason like human beings. Emotion AI is focused on training computers to recognize, quantify, and respond to human emotion, something that traditional computers were not built to do. My goal is not to build emotive computers, but to enable human beings to retain our humanity when we are in the cyber world. This book—my life—is about the quest to humanize technology before it dehumanizes us.

In striving to become the "expert" I needed to be in human emotion in order to teach machines about emotion, I found myself turning the spotlight on my own emotional life. This was an even more daunting process than writing code for computers; it forced me to confront my own reticence to share my innermost feelings, indeed, my reluctance to recognize and act on my own feelings. Ultimately, decoding myself—learning to express my own emotions and act on them—was the biggest challenge of all. Expert as I have become on the subject, I feel that I am very much a work in progress myself.

To me, my work and my personal story are inseparable; each flows into the other. And so this book is a chronicle of that dual journey—the quest to equip machines with EQ and, in the process, unlock my own EQ.

I am a rarity in the tech world: A woman in charge—a brown-skinned computer scientist at that—in a field that is still very male and very white. I was raised in the Middle East in a male-dominated culture that is still figuring out the role of women in a world that is changing with breathtaking speed. In both these cultures—tech and the Muslim Middle East—women have been excluded or restricted from positions of power. I've had to learn how to maneuver both cultures to achieve what I have.

I am who I am because I was raised by a particular set of parents, both modern and conservative; forward-thinking and yet locked in tradition. I am a Muslim, and I feel I am stronger and more centered for it; I adhere to the values of my religion even if I am not as dutiful as I once was. And I am a new American and am thriving on the energy, vitality, and entrepreneurial spirit of this great country.

I offer a perspective on growing up in my world, in a Middle East that you rarely hear about in the West. I want to introduce you to my family and share some of our experiences. After all, bridging the divide between people is how we gain empathy, and that is how we build strong, emotionally intact people and a strong, emotionally intact world. That is at the core of what I do, whether in the real world or the cyber one.

I am also passionate that people understand what AI is, and how it's going to impact their lives. The world, *your world,* is about to change. And as someone who has been at the forefront of this movement, I want to take you behind the scenes and show you how AI is built, how this technology will unfold, and, more important, how to put it to the best use. AI is no longer the flying car of some far-off future: AI is becoming mainstream. It is taking on roles that were traditionally done by humans, like driving our cars, helping us manage chronic health conditions, and even reviewing your next job application. Given how AI is becoming ubiquitous, and the potential impact it has on all of our lives, it's critical that we as a society take an active role in how this AI is designed, developed, and deployed.

With our reliance on our computers today, Emotion AI is absolutely essential. Right now, we are pushing technology to the edge, asking it to perform tasks it was never meant to address. Computers, after all, were originally designed to compute (hence the name), to crunch numbers faster and more accurately than would be humanly possible.

In this age of immersive technology, computers are being called upon to do far more than basic number crunching. Mobile technology (our smartphones, tablets, smartwatches) has ushered in a whole new world, one of presidential tweets, Facebook, Snapchat, crowdsourcing, digital banking, online shopping, and virtual assistants to perform everything from booking hotels to making stock trades to "hooking up." With devices such as Fitbit, Siri, and Alexa, we are now all connected, all the time, wherever we are.

As computers grew smaller, sleeker, more powerful (and more truly mobile), we began to use them for everything in our lives, including things for which the absence of EQ matters a great deal.

For many of us, computers have become the primary way we communicate with one another. We may all be connected all the time, wherever we are, but it doesn't mean we are actually communicating or connecting with one another in a meaningful way. We evolved over millennia to communicate face-to-face. Words alone do not convey the true meaning of a message. The vast majority of human-to-human communication is transmitted through our nonverbal signals, facial expressions, and fluctuations in voice, gesture, and body language. All of this is lost when we communicate online.

Of all our nonverbal cues, I believe the human face is our most powerful conveyer of emotion. Our faces display the full spectrum of our emotions and other mental states, from enjoyment, surprise, and fear to curiosity, boredom, love, and anger. That's why I focused my work on teaching computers how to read faces, the way humans do, so they could recognize and respond to our facial expressions, from smiles to scowls.

Decoding nonverbal cues, understanding the nuance of emotion by observing one another in real time, is something human beings do from the first moments of life. And we continue to develop this skill as we mature and are exposed to more and more people and expressions. This is how we gain wisdom; this is how we learn to be empathetic. It is an essential component of EQ.

As part of my research, I have worked closely with young adults on the autism spectrum, a complex neurological condition characterized by, among other things, difficulty recognizing, processing, and responding to the emotional cues of others. In fact, many have an aversion to making eye contact or even to looking someone directly in the face, so they miss most facial expressions. This can have dire implications for their ability to communicate with others, participate in family life, learn at school, retain a job, or sustain a long-term relationship.

Early in my work, I realized that when it comes to recognizing and interpreting feelings, computers are functionally autistic: They can't see or process emotion "data" or respond to emotion cues. By extension, I believe that when we interact in the emotion-blind cyber world, we are all rendered functionally autistic.

When you speak with someone face-to-face, you get immediate feedback by watching the other person's facial expressions and body gestures and listening to the tone of their voice when they respond to your words. If you are "neurotypical" (that is, not autistic), you are hardwired to process these emotion cues. That's how you discern the impact of your words on another person. We observe each other and react accordingly.

When we communicate via the cyber world, however, we miss the natural feedback system that enables us to modulate our behavior based on the reactions of others. The critical nonverbal portion of our communication is lost in cyberspace.

Our social media platforms, the way so many of us connect with one another, can be dehumanizing. Without any real emotional connection, it's easy to forget that we are talking to and about other human beings, and the absence of real-time social interaction twists and distorts our behavior. When it comes to the digital world, our

computers have trained us to behave as if we lived in a world dominated by autism, where none of us can read one another's emotional
cues.

I'm not suggesting that cruelty and intolerance didn't exist before
social media or that the world was a kinder and better place. Throughout human history, there have been horrific displays of empathy
deficit; genocide, mass killings, and slavery are stains on our past
(and still plague us today). What's different today though is that
with our 24/7 online world, the language of intolerance is quite literally in our faces, on our devices, all the time.

I have staked my career on the belief that Emotion AI is part of
the cure. It can help to strengthen emotional intelligence in the digital world (online, in our texts and emails, on Facebook or Snapchat
posts) and begin to repair the damage caused by more than two decades of our conducting a significant part of our lives and relationships, for the first time in human history, in an emotion-free zone.

When I began this journey more than two decades ago, there was
no Skype, FaceTime, or video conference; today all these tools are
readily available. A virtual "face-to-face" interaction is an improvement over an emotion blind one, but the reality is that most communications are still not visual. The primary form of communication
is texting—according to industry sources, the number of texts sent
annually is in the trillions. For the most part, the primary way we
communicate has zero EQ.

Some may say, "Why bother with Emotion AI? Just turn off your
phone! Stop texting! Stop tweeting! Meet face-to-face!" But, of course,
that's not going to happen. Now that the genie of texting and social
media is out of the bottle, we can't put it back.

I am a child of the computer age. I was born in 1978, when Gen X
was making way for the Millennials. Digital technology has opened
up the world for my generation and expanded our horizons. I am
grateful that I can FaceTime or WhatsApp my kids when I'm traveling on business, and can easily and inexpensively keep in touch with
my relatives halfway around the world. As a CEO working with a

dispersed workforce, I can conduct video meetings with clients and employees in London, New York, or Cairo while sitting in my office or a conference room in Boston.

My smartphone is the very first thing I reach for in the morning, as I check Twitter, my calendar, and my messages or write an email. It is the last thing I interact with at night before I go to bed. And because I keep my phone on my nightstand, if I wake up in the middle of the night, I give it a quick look. It rarely leaves my side.

But I'm no different from most of the people I know of my generation. According to a recent Pew Research Center study, 26 percent of the total population, and 39 percent of people between the ages of 18 and 29, are "almost constantly" online. By 2020, according to various industry reports, there will be between four and six connected devices for every human being on the planet. And I don't see our habits changing anytime soon. In fact, just the opposite: Our computers will become even more embedded in our lives. That is the new reality.

I'm not suggesting that face-to-face relationships are not important—just the opposite. Yes, it's unacceptable to sit at dinner texting instead of engaging in real conversation with the people around you, but the reality is that much of our interpersonal interactions today are conducted in the cyber world, and that's not going to change. (I'm a pragmatist.) So, the solution is not to turn back the clock, shut down our devices, and go back to life as it was before our computers. We depend too much on our technology to give it up. And doing so would be a terrible mistake. We need our technology now more than ever. But we need to make it smarter, better, and more *humane*. And, fortunately, we now have the tools to do this.

An Emotion AI world is a human-centric world, one where our technology has our backs and helps us become healthier, happier, and more empathetic individuals: technologies such as Google Glass, equipped with "emotion decoders" that help autistic kids better interact socially with others; semi-autonomous cars that assume control of the wheel when we're too angry, distracted, or tired to drive safely, preventing millions of accidents each year; emotion-aware devices (from smartwatches to smartphones to smart refrigerators) able to detect mental and physical ailments years before their onset; empathetic virtual assistants that can track your mood and offer timely guidance and support; human resources emotion analytic tools that can enable an HR recruiter to match the right person more precisely to the right job or team and eliminate much of the unconscious bias that occurs in hiring; and an intelligent learning system that can detect the level of engagement of a student and tailor its approach accordingly.

The potential for Emotion AI is breathtaking, but I am not naïve: When you have computers with the capacity to recognize and record the emotional states of users, of course privacy is of major concern. Emotion AI should—must—be employed only with the full knowledge and consent of the user, who should be able to opt out at any time. Emotion AI will know a lot about us: our emotional states, moods, and interactions. In the wrong hands, this information can be very damaging. That is why it is so essential for the public to be aware of what this technology is, how and where data is being collected, and to have a say in how it is to be used.

It is also imperative that AI technology be developed with all human beings in mind, that it be inclusive. Our software must reflect the real world, not just the world of an elite few. Much of my research, as you will see, has been devoted to obtaining data from a diverse population, representing all ages, genders, ethnicities, and geographical areas. If our AI fails to do this, we will be creating a new form of discrimination that will be very hard to undo, and as such technology propels us forward, we run the risk of leaving behind whole sectors of our population, and that would be disastrous for us all.

One reason I am writing this book is to provide a different and more humane vision of AI and technology. But I also hope that my story will inspire other dreamers and inventors to pursue their "crazy ideas" to change the world. All too often, we let fear derail us, especially if we are raised like I was in a risk-adverse culture. It took me a long time to believe in myself, and that held me back.

For as long as I can remember, there has been an ongoing struggle to reconcile my dreams and goals with my upbringing. The voice in my head was filled with self-doubt: It whispered, "You can't. You shouldn't. You won't." For the longest time, I listened to that voice. I stifled my emotions; I focused on doing what people around me thought was right. But the journey that has taken me from "nice Egyptian girl" to a strong "woman in charge" has been one of finding my own voice. I am no longer afraid of expressing my emotions and acting on my feelings; this gave me strength and understanding and made me a better leader, CEO, and human being. I have learned that the greater fluency and mastery we have of our own emotions, the more confident we become in allowing ourselves to be vulnerable—allowing ourselves to open up to others and with ourselves; this is how we become truly empathetic.

PART I

A Nice Egyptian Girl

1

Growing Up Egyptian

We are raising our girls to be perfect,
and we are raising our boys to be brave.

—RESHMA SAUJANI, founder, Girls Who Code

From the outside, my suburban Boston home is everything I always imagined a real American home should look like. Built in 1868, it is the quintessential New England central-hall Colonial, with a tidy fence, a brick walkway, a gray wood exterior, black shutters, a sun room, and a lovely backyard with flowerbeds.

But step inside and you can see that my home is more complex, like me. It is Egyptian American. Interspersed among the traditional round-armed sofa, Queen Anne chairs, and side tables are mementos from my place of birth, things you would find in Egyptian homes. Smack in the center of the living room, in front of the hearth, are two half-moon-shaped black silk screens with Arabic script that promises to "Protect our home from evil spirits" (a service that Alexa and Google Home don't offer). A large decorative plate perched against the wall near the kitchen door reads *Khatwa Aziza*, "Step in" or "Welcome." The entrance to the back door is guarded by a *hamsa*, a hand-shaped amulet with an eye in the middle of the palm, the ancient Middle Eastern symbol for warding off the evil eye. Just as in my mother's home, there are always scented candles burning in my home, the pungent-sweet, happy aromas of my youth: sandalwood, musk, amber mixed in with a mysterious assortment of spices. My mom used to buy her candles from a shopkeeper who made them himself; I order mine online.

Scents in particular evoke deep emotional memories. A passing whiff of perfume or aftershave from a stranger or the scent of a candle will instantly conjure up the memory of a person or place, even if I haven't seen them or been there in years. Our brain is primed to decode scent quickly; the smell center of the brain is connected to both the amygdala (the emotion center) and the hippocampus (where memories are made and stored). These are the most ancient structures of the brain, the limbic system (aka "lizard brain"), where emotions are processed.

I live 5,400 miles away from Cairo, but in my culture, no matter how far you travel, the emotional ties to the Old World remain strong.

My childhood home was a potpourri of modern and traditional values, often in conflict with one another. I was raised in a conservative environment, with strict social mores; my sisters and I were respectful, obedient, and hardworking. We deferred to our parents on all matters, big and small. Even as adults, our parents continue to exert influence over us in ways that Westerners can't understand. But I was also raised by a mother who was a trailblazer, who stepped outside the traditional role of women in the Muslim world and became one of the first female computer programmers in the Middle East. At a time when an Egyptian mother working outside the home was highly unusual, my mother held an important job with the Bank of Kuwait while raising three children. It was a remarkable accomplishment. And she set the bar equally high for her daughters.

My father also had very high expectations for my two younger sisters and me, and even envisioned us in high-powered positions as adults. In many ways, my dad trained us for this new world. At the same time, he grew up in a very conservative Egypt, deeply rooted in cultural expectations around the role of men and women in society. Inevitably, a real conflict developed between the person my parents raised me to be and the social and cultural expectations of being a nice Egyptian girl.

In hindsight, I wasn't the easiest of daughters: I was a disrupter both in my professional field as an AI scientist and tech entrepreneur in a very male-dominated industry, as well as a disruptor in my country and culture, breaking the rules of what it means to be an obedient daughter and wife. I know this put my parents in a difficult situation—yet, they never once wavered in their love or support, for which I am grateful.

I love my parents, even if we sometimes don't see eye-to-eye. This journey is one that we all embarked on together; we've evolved as a family and learned together. Regardless of how religious I was or am now, one Quranic verse has been drilled into my brain: Your parents come first. And second. And third. And those parents of obedient daughters have a ticket straight to heaven. No matter how young or old I am, religious or not, a believer or not, I want my parents to get that ticket.

In some parts of the fundamentalist Muslim world, educating girls is not a priority; in others, it can get you killed. But my parents revered education; our lives revolved around it. We were well-off, certainly, but not wealthy. My parents could have spent their incomes on fancy cars or vacation homes; instead, they used their money to foot the bill for expensive private educations for my two sisters and me, and later sent all of us to college. Whatever discretionary income they had went toward travel, so we could see the world and experience new cultures. That love of learning was ingrained in us at a young age—that and an incessant curiosity about other people and cultures.

Not everyone in my family agreed with my parents' priorities. When I was eight years old, one night during a family dinner, I overheard my uncle questioning my father's investment in our education. "Ayman, your girls will just get married, so why waste all that money on those fancy schools?" My uncle's sons, my cousins, would constantly tease my sisters and me that the likelihood of our doing anything useful with our lives was dismal.

My mother would never have challenged my uncle or my father in public, but I suspect that later that evening, my father got an earful from my mother about the value of education. She was the one who picked those "fancy schools" for us—my dad left those kinds of domestic decisions up to her. I am eternally grateful to my father for not listening to my uncle, and to both my parents for doing everything they could to instill a fierce drive within us to do whatever we wanted to do. At times, my parents must have been horrified by my choices. Getting a divorce, founding and running a precarious startup, living in the United States with my two children—this was not exactly the life my parents had pictured for me. Yet they laid the foundation that enabled me to break with Egyptian cultural norms, strike out on my own, and succeed.

I come from a close-knit, successful Muslim family, the kind you don't hear about very often in the West. My parents grew up in Heliopolis, an upscale community in the greater Cairo metropolitan area. They met at a computer-programming class taught by my father. I doubt they would have met otherwise. My mother's family hobnobbed with Cairo's elite at the exclusive Heliopolis Club on the weekends. Nor would she have met him at the disco parties she went to in college, wearing micro miniskirts and cropped tops, escorted by her older brother, Shafie.

My father, Ayman el Kaliouby, lost his father when he was five years old. He didn't have much time for fun; he had to grow up quickly. His mother was a widow with five children. My paternal grandmother never remarried, devoting her life to raising her children; they managed to get by, but the lack of an income and a father put a strain on them all.

My mother, Randa Sabry, was raised in more luxurious surroundings. Her father, Shafik—I called him Gedo, which means "grandfather" in Arabic—was in the military and was the director of Hurghada International Airport on the Red Sea, nearly three hundred miles and a five-hour drive away. My mother pretty much saw him only on holidays, but his position enabled them to live a life of upper-middle-class privilege. Her mother, Doreyya (Dodo for short), ran a complicated household, replete with a full-time cook, a housekeeper, and a driver. My mother attended an all-girls school and swam competitively throughout high school.

My parents both attended Ain Shams University, one of the top two public universities in Egypt. But my father is seven years older, and while my mother was in college, majoring in business administration and partying, he was in Austria.

As a youth, my father was a fanatical supporter of Gamal Abdel Nasser, the second president of Egypt, who led a military coup in 1952, deposing the monarchy. I think Nasser held a special appeal for my father, who saw him as a father figure. Nasser was a brash, charismatic leader, full of bravado. He nationalized many industries (including the Suez Canal), threw out foreign companies, and convinced Egyptians that they were invincible, that they could manage without any outside help. A "nationalist," he turned the country inward.

My father believed every word of Nasser's impassioned rhetoric: "Egypt ruled the world! We had the largest and strongest army in the world." According to Nasser, Egypt had the best of everything. Then came the Six-Day War in 1967 between Israel and Egypt, Jordan, and Syria, and my father's whole world came crashing down.

At sixteen, my father was too young to serve in the army. But like other Egyptians, he expected the 1967 war to be over quickly and result in a decisive win for Egypt. Yet the Egyptian Army was decimated. Much to my father's horror, his brother, who was just eighteen and didn't even know how to hold a gun properly, was sent to fight in Sinai. By then, my dad could see that Nasser's wildly optimistic pronouncements of Egypt's place in the world were not based on reality. Egypt was in serious trouble, both politically and economically.

Devastated, my dad decided that his only course of action was to leave Egypt. In college, he had majored in business and computer science. When he graduated, he'd traveled to Austria on a three-year visa, hoping to pursue an advanced degree there that would lead to a job so that he could stay. He sold newspapers on the street and took whatever other odd jobs he could get, scraping together a meager living while studying for the college admission exams. At the end of the three years, he tried to renew his visa, but the Egyptian consulate refused.

When my father returned to Egypt, he found the country more outward-looking under new president Anwar Sadat (who was later assassinated, in 1981), and his computer skills were highly valued. There was a great demand for computer classes, and a dearth of trained people to teach them. My father landed a teaching position at International Computers Limited (ICL), a British computer services company with a Cairo branch.

My father taught Introduction to Computers and programming, mainly COBOL (Common Business Oriented Language), at various universities. It was a rigorous schedule, and he worked hard. He taught a four-hour class in the morning at ICL and another four-hour class at night. Eight hours teaching a complicated subject like computer programming was intense. At the end of the day, he was exhausted. But he honed his skills and quickly became the best computer instructor in the company.

One day, he was called in to proctor an exam at a university, and a beautiful young woman there caught his eye. In her last year at university, my mother had taken an Introduction to Technology class with a friend. She was surprised to discover that she had a natural aptitude for it. She walked into the final exam feeling confident, but her friend was not as prepared; she asked my mother to share answers with her during the exam. The proctor was a tall, slim man, handsome but stern-looking. I doubt my mother would have looked at him twice, except that he was glaring at her. It was hardly love at first sight. He had caught her passing answers to her friend and took away both their exam papers. My mother was terrified that she had failed. In fact, she passed, but her friend flunked the course.

When the computer-programming craze reached Cairo, my mother signed up for an evening class at ICL for instruction in COBOL, undaunted by the fact that she was one of the few women to do so.

On the first night of class, she walked into the room and immediately recognized the instructor—it was the same tall, slim, stern professor who had caught her giving answers to her friend. She tried to switch into another class, but the director insisted that my father was the best instructor he had. So, my mother stuck with him, another sign of her ambition and determination.

My mother dreaded going to class because my dad constantly called on her. It may have been his way of flirting, because, halfway through the course, he asked her out. She turned him down, explaining that her family did not allow her to date. When my father asked her out again the next day, he received the same answer.

What do you do in the Middle East when a woman isn't allowed to date but you want to get to know her? You marry her! Like the good Egyptian boy that he was, my father followed the traditional path and asked her father for her hand in marriage.

A few months after their engagement, my father, still eager to leave Egypt, was offered a part-time teaching position at a university in Kuwait. To make ends meet, he worked at the Kuwaiti branch of NCR, a global tech company and a pioneer in online and automated banking, and did consulting on the side. For the next year, my parents conducted a long-distance relationship, by letter.

They were married in July 1977. Their wedding portrait shows my father, a strikingly good-looking man, in a crisp white suit, and my mother in a white lace dress, her dark hair framing her face and cascading to her shoulders. And standing behind them is one of Egypt's most famous belly dancers, wearing the traditional *badlah,* a tight-fitting top, jewel belt, and long, sheer skirt. (Belly dancers were common at Egyptian weddings then.)

My mother joined my father in Kuwait and was immediately hired at the National Bank of Kuwait as a computer analyst. I was born in August 1978. My mother then did something very unusual, even bold, for a Middle Eastern woman: She continued to work outside the home after I was born.

In the Middle East, a woman is not expected to take her husband's name after she is married, so my mother kept her family name, Sabry. For children, however, your full name is your given name combined with your father's full name. So, my official name is Rana Ayman el Kaliouby.

As a nice *married* Egyptian girl, my mother had to follow my father's fashion dictates. As he came from a more conservative home, her wedding dress was the last form-fitting dress she ever wore in public. From that point on, she turned in her miniskirts for slacks and modest-length skirts. She did not wear bathing suits on public beaches, and she kept her arms covered. She did not, however, wear a hijab, the traditional head cover for modesty—that would come later, when I decided to wear one.

2

Oil and Water

I've always thought of myself as an Egyptian, even though I spent the first twelve years of my life in Kuwait.

Kuwait is not like the United States: In Kuwait, if you're not Kuwaiti, you are "other"; you are an expat always. There is no path to citizenship. And if your mother or father isn't a Kuwaiti citizen, it doesn't matter if you were born there; there is no birthright citizenship.

While we were not treated particularly badly, I can't say that the Kuwaitis rolled out the welcome mat for us, either. For the most part, expats and Kuwaitis didn't mix. We were like oil and water. Many Kuwaitis were far wealthier than the expats. They lived in fabulous villas on the Gulf, drove luxurious cars (most had drivers), and kept servants, mostly from Indonesia or the Philippines. We did not travel in the same social circles.

But I loved our years in Kuwait. If you didn't mind being a perpetual outsider, and you were one of the skilled, better-paid workers, you could enjoy a nice life there with the other outsiders.

My parents had been attracted to Kuwait by the bigger paychecks and a middle-class lifestyle. Kuwait sits on the sixth-largest oil reserve in the world. But it didn't have enough skilled Kuwaitis to fill positions in finance, healthcare, office work, education, technology, or the other occupations that keep an economy and a society humming.

So, flush with money, Kuwait had imported its brainpower, luring skilled, educated people from all over the world, especially from elsewhere in the Middle East, and from India, Europe, the Philippines, and North Africa. Faced with a sluggish economy at home, educated Egyptians were particularly attracted to Kuwait. Hundreds of thousands of Egyptians immigrated there; some half a million live there now.

The Egyptian expats in Kuwait pretty much stuck together. Since only Kuwaitis were legally allowed to buy property, expats lived in rental apartments in complexes that catered to foreigners. Our last apartment, in Salmiya, had four bedrooms and a spacious foyer that led into a family room with huge windows overlooking the Gulf.

My father's one personal indulgence, not surprisingly, was technology; he was always an early adopter. We had a video camera and a VHS player long before anyone else, which made him the official documentarian of family events. My earliest memory in Kuwait is of a child-size plastic blue chair that I called my royal blue chair, my throne. When I was around five, I'd stand on top of the chair and I would just talk, giving one speech after another about anything that popped into my head. My dad videoed my talks and gave me tips on public speaking ("Rana, look at the audience; enunciate your words!"). This was my first exposure to technology, but more important, it was my first experience speaking before an audience, even if it was just an audience of one. This was my special time together with my dad, and I loved getting all his attention.

We were also the first among our set to have an Atari 2600 computer, an old-school console with a slot where you inserted game cartridges. Never one to squander a teachable moment, my father made us figure out how to set it up. Perhaps that explains my ease with computers. I have never been daunted by having to build a piece of technology from scratch. If you persist, you can eventually figure it out.

Still, I was not the stereotypical nerd transfixed in front of a digital screen. As my family members and I were wiping out alien invaders, we would talk and catch up with one another. To me, technology was a social tool, a platform to bring people together.

From kindergarten to third grade, I attended the Sunshine School, which followed a British-approved curriculum. Classes were taught in English (with the exception of Arabic-language classes), with an emphasis on the liberal arts, music, and sports. In contrast, most Middle Eastern schools focus on rote education.

The Sunshine School was founded by Vera Al-Mutawa, a proper but warm English lady who was married to a Kuwaiti sheikh. She had some very progressive ideas about education. I'm not sure I would have made it past kindergarten without Mrs. Al-Mutawa's intervention. You see, I am left-handed, and according to Islamic tradition, you should eat and work only with your *right* hand. That makes anyone who is a lefty, like me, a bit of a pariah. So I spent much of my early years trying, with little success, to sublimate my natural impulse to use my left hand. Mrs. Al-Mutawa noticed that I was struggling to hold a pencil. Familiar with Arabic culture, she quickly realized that I was a lefty being "trained" to be right-handed. She contacted my parents, who were surprised to learn that "handedness" is encoded in your genes; people really don't have a choice in the matter. My parents may be traditional in outlook, but they also respect science: From that point on, they no longer tried to cure my left-handedness, and I began to excel in school. By the summer before first grade, I was so far ahead of the rest of my class that Ms. Al-Mutawa recommended that I skip first grade altogether and go directly into second grade. And so for the rest of my academic career, I was the youngest in my class by a couple of years.

. . .

I may have been sent to Western-style, more progressive academies than the typical Middle Eastern student, but at home, traditional values reigned, especially when it came to the role of men and women. On the one hand, my father was extraordinarily progressive for the Middle East: He believed that women should be educated and he himself married a successful, educated woman. He was very supportive in encouraging us to do our best and make something of ourselves. On the other hand, like most Egyptian men—and I mean most Middle Eastern men—he expected my mother to wait on him hand and foot. For example, I never saw my father walk into the kitchen to get a glass of water. He would ask my mother to get it for him, and she would comply. I don't even think he realized what he was doing; it was ingrained in how he was raised.

As my sisters and I got older, my father began asking us to fetch things for him. I never questioned this, until I was around nine years old. My father and I were sitting in the den, and he said to me, "Ranoon [that's my nickname], get my black shoe shine kit from my closet!"

I was annoyed; my first impulse was to say, "Go get it yourself!" Of course, I would never talk back to my father—it was shocking enough that I even thought that way. But I was incensed that he expected me to fetch things for him. So I devised a strategy of passive resistance—I played dumb. I dutifully went into the bedroom, opened his closet, and deliberately picked the brown kit, knowing that he wanted the black. I brought it to him with a smile: "Here, Pappy." He looked at me, shook his head. "That's brown. Go get the black." I went back and shuffled through the closet, taking my time until I brought him the right one.

I practiced this form of silent rebellion often, pretending to misunderstand what he wanted or just bringing the wrong thing. Eventually, he stopped asking me to get things for him and turned to my sister Rasha.

I don't want to make it sound like I walked around my home seething with resentment—I didn't. But as I watched my mother cater to my father, and my aunts cater to my uncles, and my grand-mother to my grandfather—the list goes on and on—I remember thinking, "I don't want to marry a typical Egyptian man!"

THE HIJAB

In the Middle East, what a woman wears is not just a sign of her fashion sense or her devoutness, but also a reflection of the cultural and social trends of the day. At times, some of the women in my family were "veiled": They covered their hair and necks with a scarf or a turban when outside their home, and so did I. My sister Rula and I are now unveiled, while my mother wears a hijab, as does Rasha, my middle sister. A hijab frames the face and, without the distraction of hair or jewelry, accentuates the eyes, which are argu-ably our most expressive feature. People think of a smile as being all about the mouth, but without those crinkly smile lines around the eyes, a smile is only half-hearted, or can even be fake. My mother's older sister now wears a niqab—she is covered from head to toe, with just a small slit left for her eyes—and although she is fully covered, I can tell whether she's had a good day or a bad day just by looking into her eyes.

Every summer, my parents would take a month off and we'd spend two weeks visiting family in Cairo; one week at a time-share in Alexandria, the second largest city in Egypt, with beautiful beaches on the Mediterranean; and one week in Europe—we alternated among Cyprus, Spain, Belgium, Switzerland, and Nice/Monte Carlo.

Perhaps my favorite part about these trips to Europe was plan-ning for them. My dad had invested in the Encyclopedia Britannica, which was quite expensive at that time and included a huge atlas. Every summer, as my dad planned which country we would visit, he and I would lay out the atlas on our dinner table and he'd help me trace with my little fingers where the different cities were. He instilled in me a sense of wonder about the world. A sense of adventure. There was no fear of the "other," only curiosity and open-mindedness. This is how I got my love of travel, but also how I ended up an open-minded citizen of the world.

And my parents, for all their conservative values, did not censor what we saw. In the south of France, every woman on the beach was wearing a skimpy bikini, and many went without tops. In contrast, there was my mother, sitting quietly on a blanket in the sand, tucked under an umbrella, covered from her neck on down in a loose-fitting two-piece outfit. We understood that our traditions may be different, but we weren't shut off from the rest of the world.

Kuwait can get oppressively hot in the summer, but my family was lucky enough to escape the heat at our private beach club for Egyptian expats on the gulf. My favorite time of day was sundown; the air was fresher, and a slight breeze would pick up off the water. When it was too hot to cook at home, we would have family barbecues on the beach with my uncle and his family. When dinner was finished, my sisters, my cousins, and I would race over to an ice-cream cart owned by a vendor on the beach from whom we'd buy the best ice-cream pops in the world. They were vanilla inside, encased in a hard raspberry-flavored shell—creamy, not too sweet, not too tart, just right. And on those evenings on the beach, it seemed like life was perfect.

I had no idea that the Middle East was on the brink of another war. I was worried about the kinds of things that concern eleven-year-old girls. I remember sitting on a picnic table in the schoolyard at the end of seventh grade, fretting over a particularly difficult math assignment, with my best friends Nisreen, Hanan, Rania, and Yasmine. I was looking forward to returning in the fall and picking up with my friends where we had left off.

I didn't realize that our life in Kuwait was about to change forever.

3

Uprooted

Emotion is universal. We are born in different countries, practice different religions or no religion at all, and live very different lives, but our emotions unite us. People of every background and zip code draw upon the same emotional palette: joy, fear, anger, disgust, love, hate. Nevertheless, there are significant cultural, ethnic, and even gender differences in how we *reveal* our emotional states to others.

In some cultures, you are criticized for being too emotive, or for exhibiting the wrong emotions in front of the wrong people. For example, based on my company's research, in China and India, where group goals supersede those of the individual, people are likely to dampen or mask their emotions in the presence of strangers, especially negative emotions such as anger and contempt. Those emotions are considered self-indulgent. Women around the world smile more often than men—we are socialized to please—but younger women smile a lot more than older women. The exception to the rule is Great Britain, where men and women smile about the same amount.

Egyptians are one of the most expressive, *emotive* people in the world. In a way, my education in the science of emotions began on my family's annual visits to Cairo for summer vacation, sitting around my grandmother's dining-room table, surrounded by my parents and sisters and two dozen or so aunts, uncles, and cousins. I watched with fascination as members of my big and warm extended family talked (all at the same time), gestured with their hands, laughed out loud without restraint, interrupted one another, and engaged in lively, spirited conversation and debate, amid heaping plates of food.

Looking back, I see that it was at my grandmother's that I began to notice the cultural differences in how emotion is expressed, a fact that I took into account only later, when I was designing software that would have to read and interpret our emotion cues accurately, whether it was observing someone who was Eastern or Western, male or female, young or old, emotive or reserved.

My grandparents lived in a 1950s development of several dozen vanilla-colored, oblong low-rise concrete structures. Each floor had a terrace with horizontal metal railings. Their house was originally built with two floors, like most of the other homes in the neighborhood, but my grandparents eventually "popped the top" and added three additional stories, creating a flat for each of their grown children. The backyard is a lush garden—it was my grandmother's pride and joy—a fresh, green oasis among the sand and dust of a country that is 90 percent desert, offering relief from the sizzling hot summer temperatures that often top 100 degrees Fahrenheit.

My grandmother Dodo grew flower palm trees for precious shade, grapevines for their delectable grape leaves, guava plants for the medicinal value of their leaves (an old-time remedy for digestive problems), and the beloved mango trees for their luscious fruit. There are at least ten different varieties of mangos in Egypt, each with its own unique properties, all succulent and bursting with flavor (unlike the yellow, hard, odorless mangos sold in the States). My favorite is the Ewesi mango, sweet and fragrant, reddish on the outside, soft and golden on the inside. Our summer vacations coincided with peak mango season, when the Ewesis were literally dropping off the trees in my grandmother's garden.

The minute we landed in Cairo, we'd drive directly to Dodo's villa, where we'd make a beeline to the back-door stairs leading to the kitchen. Dodo would be sitting at a table in the middle of the kitchen in her turban—my grandmother covered her head even indoors, for modesty—slicing, dicing, stuffing, and giving detailed instructions to the two maids helping with the cooking for our family's annual reunion. Her kind face would light up with a big smile when we hugged her.

In the summer in Egypt, the big meal of the day is served at around four P.M. After dinner, we would race into the main hall, where the huge table there would be covered with old newspapers and about a hundred mangos picked from my grandma's garden. The smell was incredible: a bouquet of citrus and peach, very intense and sweet. Strong emotional responses in our lives help lock in memories. That is how, decades later, I can recall not only every dish my grandmother served, but also the scent of the food. At her house, I felt loved and safe.

THE INVASION

My family and I were visiting Dodo and Gedo in late July 1990. My dad spent a few days with us before he had to go back to work in Kuwait; my mother, my sisters, and I stayed on for an extra week.

On August 2 at two A.M., while we were all sound asleep, Saddam Hussein, then president of Iraq, invaded Kuwait. Within two days, the Kuwaiti government had fallen, and the country was under Iraqi control.

We huddled together on my grandmother's king-size bed, watching the images on television of tanks rolling through neighborhoods we knew as home. We were frightened by the reports of Iraqi troops storming from house to house, seizing or destroying property. My mother tried to call my father, but the Iraqis had cut all lines of communication. For two weeks, we didn't know whether he was dead or alive. My mother kept assuring us that he knew how to take care of himself. We believed her because we wanted to, but we understood more than we let on.

Being uprooted unleashed in me a flood of emotions I had never experienced before. I didn't realize at the time how angry I was at the situation, how frightened and uncertain about the future.

I could never talk about my fears or concerns, not to my parents, or even to my sisters. It wasn't part of our family's ethos. My family lived by a code: work hard, stay focused, always give whatever you do your best efforts. If you encountered an obstacle, you rallied to overcome it. Expressing negative feelings would have been seen as whining or complaining—that was unacceptable. Instead, my family took the practical "let's tackle this problem and solve it" approach. We didn't look back; we forged ahead. As the oldest daughter, I felt a particular responsibility to remain strong and hold my emotions together. I didn't want to disappoint my parents or let my sisters down. I put a great deal of pressure on myself, and looking back, I realize that I shut down emotionally. Rather than deal with the fear and anxiety I had around the situation in Kuwait, I pretended it didn't exist.

For a while, I clung to the wildly optimistic belief that we would eventually go back to Kuwait to resume our lives—anything else was unthinkable. My mother was more realistic. Every day, riveted to the news, she saw the life she and my father had worked so hard to build going up in smoke.

In the blink of an eye, both my parents had lost their jobs, their savings, and their home. Yet we were better off than most expats: We had a loving family that took us in and a roof over our heads. We moved into one of the empty flats in my grandparents' villa along with our Filipino nanny, Linda, who had accompanied us on the trip.

My grandfather assured my mother that Iraq would eventually be defeated. But he also knew that this ordeal would not be resolved quickly, and that there was no chance of our going home anytime soon. My grandfather advised my mother to start looking for schools for us. That's when it began to sink in that our lives were never going to be the same.

Years later, my mother confessed to me that during this time she was deeply distressed over my father's absence and their prospects for the future. But she hid it well from us. There was much to be done—the beginning of the new school year was fast approaching. My mother went on a whirlwind tour of Cairo's top private schools. Her goal was to duplicate the progressive but academically rigorous, British-style liberal arts education we'd had at our old school. Finally, she found a school that met her high standards: the Thebes International School, a respected private school that was opening a new campus in Heliopolis, close to where we lived. I think what sold her on the school was the Olympic-size swimming pool, which meant that my sisters and I could continue swimming competitively.

Then, just as school started, we finally received some good news. My father called. He had been hiding in the apartment since the invasion. He was told that foreign nationals would be allowed to leave Kuwait soon. A great weight was lifted in our household. In early November, he and a few friends were allowed to leave, driving a long and treacherous route through the desert from Kuwait via Jordan to Cairo.

When my dad returned, he started working in Cairo as an IT consultant. Soon, my mother was back at work, too. The National Bank of Kuwait had set up shop in Cairo. They paid my mom a percentage of the monthly salary she had received in Kuwait, to help keep the bank operational, which was a real lifesaver.

The Thebes School, while academically strong, was much more traditional than my school back in Kuwait, with a greater emphasis on rote learning and memorization. I had a reputation for being a studious, well-behaved student, so the headmaster chose me to be class prefect, which is sort of a glorified hall monitor. In some classrooms, even the most minor infractions would elicit a rapid and harsh corporal punishment. One day, I showed up in my homeroom class wearing a blue hair band that was a few shades lighter than the officially approved school color. The irate teacher walked over to my desk, her face red and furious, and abruptly slammed a ruler down on my hands. I was outraged by her assault, but I kept my mouth shut. This was the school's culture. From then on, my mother and I did a careful check of my headband every morning before school.

People unfamiliar with the Middle East often think of it as one big monolithic entity, where people dress the same, eat the same food, and follow the same customs and rules. Nothing could be further from the truth. Each country has its own distinct personality and cultural norms. My Egyptian school may not have been as academically progressive as my Kuwaiti school, but socially, Cairo was light-years ahead of the more conservative Gulf countries. Many teenagers at my school were allowed to date. Compared to my sedate, buttoned-down Kuwaiti school, Thebes was loud and chaotic. Kids talked back to teachers. Boys and girls held hands in the yard. A few even smoked. I was shocked—that was way out of my comfort zone.

All this was quite eye-opening, and very unsettling.

I was the new kid on the block, raised in a different country with different cultural mores, and because I had skipped two grades (first and eighth), I was just twelve years old, two years younger than everybody in my class. And given my parents' no-dating policy, I was doomed to be an outsider in the school's social scene, too. This made me feel even more isolated from my classmates, but I was also fascinated by this new environment. I became very curious about the intricacies of boy-girl relationships. So, I began studying my fellow classmates.

I began observing their faces, watching the boys and girls in class exchange furtive looks, and monitoring their actions after class. I began to figure out who liked whom, and who was on the verge of breaking up—often before the couple themselves knew it.

For example, I noticed that Rashida, a pretty, dark-eyed girl with long, wavy hair who was dating Mukhtar, a gifted science student, was constantly casting glances at Mohammed, a good-looking boy with an athletic build. Mohammed, in turn, would look back at her and then quickly look away. A few days after I first noticed this, Rashida was holding Mohammed's hand, while Mukhtar looked forlornly at his now ex-girlfriend. So, I had an uncanny knack for getting it right, and eventually I felt confident enough to make predictions to my friends about their future romances: "Oh, he has a crush on you," "Watch out, he's got a roving eye." Before long, I became the go-to expert for romantic advice.

I made a few friends, but I still felt very lonely. I was something of a misfit, not just because I couldn't date, but because, unlike most teenagers, I loved school and I loved studying. This really set me apart from my classmates. I didn't complain about homework; I looked forward to it! Long after the household had gone to sleep, I would stay up late at night, solving math problems, reading, thinking, with books sprawled all over the dining-room table. On one of these nights, I looked out the window and noticed that the neighborhood was dark, except for a single light two buildings away. There in the distance I could see a young man sitting at a desk, with a lamp, absorbed in a book, studying, too. He looked up. I waved and he waved back. We would stay up late and synchronize when we'd switch off our lights. After midnight, one of us would take the lead and call it a day and the other would follow. It was almost my way of flirting. I felt like I was in a relationship, and he was my secret late-night date. In my naïve young brain it was a borderline romantic relationship—my first boyfriend, albeit from a distance.

On February 28, 1991, Kuwait was liberated by Operation Desert Storm, and the Gulf War officially ended—but the chaos was far from over. In a particularly vicious act, Saddam's departing soldiers set the Kuwaiti oil fields on fire, creating an environmental disaster in the region. The fires burned for almost a year, the hot, putrid oil spewing into the Persian Gulf. Soon after, the National Bank of Kuwait hired my mother back full-time, and the Kuwaiti ministry hired my father back. My parents were on the first private plane back to Kuwait. They left us with our grandparents. When my mom said goodbye, it was the first time I had seen her cry since we were uprooted.

I felt scared. I was worried about what would become of us. How would we survive this ordeal without my mother to help us through it? She was our rock, the person we depended on; she was holding us together. Of course we were left in kind and competent hands—my grandparents were wonderful to us, as was my nanny, Linda, who had worked for us for several years. The sight of my mother boarding that plane in the midst of all that disruption was devastating. Yet I didn't cry; I kept my feelings to myself.

For the most part, I was a master at suppressing my feelings. I behaved as if nothing had happened. But sometimes my emotions bubbled to the surface in unexpected ways. I lashed out verbally and in one case physically. In one instance, I punched a schoolyard bully! Everyone at school was stunned because I was such a Goody Two-shoes! Another time, I became enraged when I realized Linda had borrowed my hair rollers without telling me. It wasn't a big deal, yet I went ballistic, screaming at her with such ferocity that she started to cry. I felt terrible. I knew my mother would have been furious at me if she had witnessed my behavior. This was not acceptable in our home. At the time, I couldn't understand why I blew up the way I did. Now, of course, I realize that I was acting out.

Being uprooted changed my life in ways I didn't fully appreciate. It made me tougher and more determined to excel, unleashing my sense of ambition and competitiveness. I was determined not to disappear into the sea of unfamiliar faces at my new school. I wanted to stand out, to make my mark. I couldn't change the fact that the war had disrupted our lives or that I was in a school that made me feel like a geeky outsider. So I focused on what I could control; I channeled all of my energy into my schoolwork. This became a lifelong pattern, a way of coping when I felt overwhelmed. I buried myself in my books. I drove myself hard to succeed, and by the end of the year, I was at the top of my class.

When school ended in late spring, we visited my parents in Kuwait. It was the first time my sisters and I traveled on a plane without our parents. We sat three-year-old Rula in the middle between Rasha and me. I was scared, but I refused to show it. I tried to make the flight to Kuwait fun for my two sisters. I felt a rush of relief when we got off the plane and saw my parents smiling and waving, and my mother running over to hug us. I could finally exhale.

They were living in a new apartment; our old apartment had been completely ransacked by the Iraqi Army. Our plan was to move back to Kuwait as a family, but the oil wells were still burning, and the air was so dark and sooty that you couldn't tell day from night. Once, when my sisters and I ventured out to a nearby playground, we returned home covered in black ash. I can still remember the stench of the burning oil, the acrid taste of soot in my mouth. I had to scrub it from under my fingernails and wash it out of my hair.

For my parents, this was the final straw. We left Kuwait for good shortly afterward and moved back to Egypt. My sisters and I were sent to live with my grandparents again, for the new school year, while my parents looked for jobs.

My father obtained a position with the government of the United Arab Emirates that was similar in salary and status to his job in Kuwait; my mother was hired there as a teacher of computer science at a local school. And so, we moved to Abu Dhabi.

4

What Will the Neighbors Think?

n Egypt, you begin applying for college the summer after your junior year of high school. Because I had skipped two grades, I was only thirteen when I began that process.

Nice Egyptian girls don't live on their own, not even in an all-girls college dorm, so I was restricted to applying to schools in Cairo, where I could stay with my grandparents. My parents allowed my sisters and me to choose from among three majors: computer science, medicine, and engineering.

I had always been drawn to people, did well in science, and wanted to do something that was important. Perhaps that's why I first flirted with the idea of enrolling in a premed program and becoming a doctor—what's more important than saving lives? Still, I was equally intrigued by the prospect of studying computer science, which might seem an odd choice for a "people person." But to me, technology had always been more about the people than machines, perhaps because playing videogames had been a way my sisters and I connected with one another. I also believed that computer science was the career of the future, and that's why I found myself increasingly pulled in that direction, too.

I had been a straight-A student, and applied to the two top schools in Cairo, the American University of Cairo (AUC) and the more affordable, highly rated public university. The public university was a typical Middle Eastern school. Most of the professors were Arab, the classes were taught in Arabic, and as a premed major, I would take only science courses. AUC was more like the British academies I was used to. Half the professors were American, the classes were taught in English, and the curriculum was a traditional liberal arts program, even for science majors.

When I was younger, my life revolved around my Muslim faith. My family routinely went to mosque on Friday afternoons, the day of the Muslim Sabbath. For a while when I was a young adult, I would often excuse myself from work, find a quiet spot to put down my prayer rug, and pray five times a day: before sunrise, midafternoon, late afternoon, sunset, and at night. When I had to make a difficult decision like this, I often turned to prayer, as I do now. There is a special prayer in Islam called *Salat-l-istikhara,* in which you ask Allah to guide you in the right direction when you are unable to make up your mind. A rough translation of the prayer is "Dear Allah, you know what's best for me, so please lead me that way."

I don't think I ever got a clear sign of what to do, but after I recited the prayer, I would experience a sense of relief that things would happen the way they're supposed to and that everything would work out, because that's the way it was meant to be.

I was accepted by both schools, but I soon realized that I didn't have the calling to be a doctor. I also realized that AUC was a better fit for me. So, I set my heart on AUC, where I intended to study computer science.

Still reeling from the financial blow inflicted by the Gulf War, my parents not only were now facing my college tuition, but were still on the hook for my two sisters' private-school tuition. For the first time, my father balked at the expense. He felt I should go to the less expensive public university, where he and my mother had gone.

We applied to AUC for financial aid, but the amount of student aid available was very limited. Then my mother intervened, forcefully. "Education is the best investment," she declared. Even if it took her entire paycheck for the next four years, she said, I was going to go to the best school I could get into. She never wavered in her support. I think her resolve surprised my father, who eventually gave in. Then, to my parents' relief, and mine, I was awarded a generous merit scholarship to AUC. My prayers had been answered.

As a fifteen-year-old freshman at AUC, I lived with Dodo and Gedo. They were warm, kind, and supportive, but they imposed the same rules as my parents. I was not allowed to go to parties, even those on campus; I had a strict curfew; I was forbidden to give out my home phone number to any of my male classmates; and I was not allowed to date. I never questioned or challenged them. I knew my parents and grandparents trusted me, and I felt honor bound never to break that trust.

There was a great deal of wealth in Cairo, especially in the nearby suburbs and in the posh gated communities. But there were also deep pockets of poverty in the inner city, especially in the area surrounding Tahrir Square, where at the time the main AUC campus was located. For many of the residents, particularly the younger ones, their condition seemed hopeless. It's very difficult to gain entry into the middle-class in Egypt and throughout the Middle East.

AUC was a half-hour drive and a quick subway ride from Heliopolis. Whatever time of day you rode the subway, the cars would be bursting at the seams with passengers. Rush hour was particularly bad.

When I emerged from the subway, I would run the gauntlet of men of all ages who hung out in the square and who felt it perfectly appropriate to make sexually suggestive remarks to me, an unescorted fifteen-year-old girl. I would walk briskly, head forward, avoiding eye contact and keeping a neutral expression on my face. I felt vulnerable, my heart beating fast. When I finally reached the gates of AUC, I would breathe a sigh of relief.

To the denizens of Tahrir Square, AUC was just an extension of Western corruption. To make matters worse, more than a few wealthy students and faculty drove in and out of the campus in Mercedes-Benzes, BMWs, and Porsches, which only added fuel to the fire.

Inside the gates, the grounds were green and lush, with palm trees and gardens. The century-old buildings represented the best of Arabesque architecture, with arched windows and elaborately carved ceilings.

Although I was majoring in computer science, I was required to take courses outside my field. *But what does English literature have to do with Bayesian networks, CPUs, bytes, and Baud rates?* I thought. Still, those classes opened my mind to new ways of thinking and looking at the world. And Economics 101, Organizational Behavior, and Marketing 101 gave me insight into how people think and make decisions.

At AUC, there were equal numbers of male and female computer science majors, and this is true throughout the Middle East. Indeed, I was shocked to learn that in the United States, computer science majors, and majors in all other STEM (Science, Technology, Engineering, and Math) disciplines, are overwhelmingly male. In the Middle East, women actually outperform men in STEM subjects, perhaps because we have to work harder to prove ourselves.

The fact that I was forbidden to date or even attend organized school events like co-ed concerts locked me out of the social scene and solidified my geek outsider status. Still, I developed some close friendships with women and became the unofficial class "confidante." My friends poured their hearts out to me about their boyfriends and breakups, and even divulged secrets about their families. I was a good listener and kept their confidences; but the reality is, it was a one-way street—I confided little in them. I didn't feel comfortable sharing my inner thoughts (and I still feel uncomfortable doing so to this day). I had nothing to contribute about my own love life. I had none! I did have crushes on some of the guys at school, but I never breathed a word about it to anyone. I knew that it was pointless to have these feelings, since I could never act on them. I bought into my parents' belief that dating would be a distraction from my real purpose for being at school, getting my degree. I never questioned my parents about this; it was just the way our family rolled.

True to form, I laser focused on my schoolwork. It turned out that computer science was the right choice for me; I loved my classes and excelled at writing code, or "coding." To non-techies, coding may sound like a lot of monotonous number crunching, but it can be very creative, too. Source code (or just "code") is a blueprint, a set of instructions that provides a computer with information to complete a task. Everything you do on a computer, from sending an email to running calculations to playing a song to controlling the thermostat in your home, requires a set of instructions written in a language the computer understands. Just as there are numerous spoken languages, there are many programming languages, with names like Java, Python, C++, JavaScript, Ruby on Rails, and Perl.

All spoken languages conform to a basic grammatical structure; for example, every complete sentence has a noun and a verb, and whether we are writing or speaking, we build upon this model. Once you master the structure of one language, you have the basis to learn other languages. The same is true of computer languages. Once you understand the basic structure, with time and practice you can easily learn new languages. Writing code is as much an art as it is a science; good code is elegant and efficient. Well-written code tells a story that is easy to follow. As with a well-designed map, the instructions are clear, precise, and get you from point A to point B as quickly and efficiently as possible. In programming lingo, good code is "well documented," meaning that if another programmer looks at it, it should be evident to her what the code is for. Poorly written code, on the other hand, is overly complicated and just plain sloppy. Programmers refer to it as "spaghetti code," because the information is tangled up like strands of spaghetti. Have you ever downloaded an app onto your smartphone only to find that it gobbles up so much of your battery power that you have to delete it? This is an example of poorly written code; any coder worth his salt knows that you can't create an application (a program) that makes it impossible to run other vital functions on the same platform. Similarly, apps that are "buggy" (i.e., that have a fatal error or flaw) tend to crash often.

MY ALL-NIGHTER

When I was in college, laptop computers were expensive and rare. They were also not powerful enough to do the kind of programming work we needed to do. So, computer-science students were tethered to the lab and its servers; the cloud did not exist back then. We were often assigned team projects, and we would meet in the evening after classes, either at the lab or at one another's houses, to do follow-up work. If I was assigned to work on a co-ed team, as was usually the case, my grandparents insisted on meeting the team. If possible, they wanted the team to meet at our home, where we could be supervised.

I spent a great deal of time writing code. I programmed a traffic-light system and a graphics UI (user interface), both incredibly useful in terms of learning how to write well-documented code. The highlight of my college experience was Course 492, a two-semester project in which students researched, designed, and launched a program that re-created a real-world work situation. My team, which consisted of my best friend, Alia, and two male students both named Mohammed, was assigned to write a program to run an elevator. On the surface, that may sound simple. But elevators are very complex pieces of machinery that perform complicated tasks.

In a perfect world, when you press an elevator call button, you expect the elevator to come within a few seconds and deliver you to your destination as quickly as possible, with no unnecessary stops or mishaps (like getting stuck between floors). You want a smooth, quick, safe ride. But if you live and work in a multi-floor dwelling, you know that there are times during the day when the elevators are full of people: early morning, when people are rushing to work, or dinner hour, when they are returning home.

My team had to create simulations for every conceivable scenario. For example, what if a passenger gets on at the fifth floor and wants to go to the twentieth floor, but another passenger gets on at the twenty-first floor and wants to go down to the lobby; or a passenger wants to go from the fifth floor to the eighth; and so on? We also had to program in a sense of "fairness"—the program has to take into account waiting times. And if there is more than one elevator, each has to be synchronized.

We spent weeks interviewing building managers, energy company executives, and business and residential tenants to assess their needs and get their input. And only after we had gathered the data on how people use elevators, and the expectations of the owners of buildings, were we able to begin creating the simulations and writing the code.

Sometimes, to meet a deadline, the team would have to pull an all-nighter. After we were done with our classes during the day, we would gather at the computer lab, working from six P.M. to four A.M. on many nights. Despite the long hours, we shared an incredible energy and camaraderie. We would catch a few hours' sleep at home, go back to class in the morning, and then regroup at the lab to start all over again.

By late fall, we hit a particularly intense work period, one that required several all-nighters in a row; graduation was in February, which meant we had three months or less to wrap everything up. I still couldn't drive, so we were grateful that one of the Mohammeds drove us home after those late nights at the lab. I lived farther away from the school than Alia, so she would be dropped off first.

During one of our hectic weeks, my father was in town on business, staying in a hotel in downtown Cairo, close to his meetings. We weren't able to spend much time together, given our hectic schedules, but we carved out some time to have lunch at his hotel. The night before our lunch, I was dropped off at my grandparents' place at five A.M. I was so tired that I fell asleep in my clothes. I got up when the alarm went off a few hours later, showered, put on some clean clothes, and went back to school.

As tired as I was from the marathon coding sessions, I was excited about seeing my father and sharing the progress my team had been making. I couldn't wait for my morning classes to end. The plan was for my father to visit with my grandmother in Heliopolis before picking me up on campus and driving us to his hotel downtown for lunch. At one P.M., I walked over to the campus parking lot and saw my father waiting for me in his red Mazda. I slid into the passenger seat expecting a warm greeting, a hug, and a smile. Instead, I got a frosty stare.

"Rana," he shouted out in anger, "*Mashya 3ala 7al sha3rik!*" (The numbers represent the sounds in Arabic that don't exist in the English alphabet.)

It was about the worst insult you could hurl at a nice Egyptian girl. It roughly translates to "the walk of shame."

"The neighbors saw you stepping out of a guy's car at five A.M.! What do you think that looks like?"

Apparently, a "well-meaning" neighbor had seized upon the opportunity to report on my comings and goings. That was all my father had to hear, that his daughter was the subject of salacious gossip. It didn't matter that he knew I wasn't out partying. What mattered was what the neighbors thought. I was in tears. I defended myself, telling him that the all-nighters were a requirement of the major. If my team didn't do it, we would fall behind.

"Well, if this is the only way to do this project, then you have to switch majors and start all over again," he said.

"What? What should I study?" I asked in shock, tears streaming down my cheeks.

"Rana, you can study accounting. I'm sure that those students don't have to stay up all night."

I was thrown into a panic: I was at the top of my class, one semester away from graduation, and my father was telling me to throw it all away and study something that I had no interest in. Was my life about to be turned upside down because of one nosy neighbor?

We never did have lunch; my father drove me home. I felt drained. Over the next few days, I was on pins and needles during intense negotiations between my father, who was determined to salvage my tarnished reputation, and my mother and grandfather, who understood how hard I was working, and the unfairness of the situation.

I don't think that my father really expected me to switch to accounting. But I know he felt humiliated by the fact that the neighbors were talking, and that this reflected poorly on his parenting. And he wanted to protect me from being the subject of vicious gossip.

Eventually, he relented: I could remain a computer science major as long as I abided by a strict curfew of eleven P.M. My team members understood my family situation and therefore delivered me promptly to my doorstep by eleven on late nights, readjusting their schedules so that we wouldn't have to pull all-nighters. Somehow we made it work.

Two months later, all the competing teams demonstrated their senior project in front of the entire computer science department and invited guests. Our presentation, and our software, was flawless.

My father was sitting in the back, beaming with pride. After the presentation, he rushed over to me and gave me a big hug. I felt vindicated. We never spoke about the incident again.

On Thursday, February 12, 1998, before a packed auditorium, three hundred graduates in black caps and gowns filed into the graduation hall. After we settled down, the vice chancellor announced, "Mr. President, the president's cup is presented at each commencement ceremony to the student who ranks highest in the graduating class. This year's winner is . . . Miss Rana Ayman el Kaliouby."

My hard work had paid off. I walked up to the stage, and the chancellor handed me a large silver cup with my name engraved on a small plaque at the bottom, along with the names of past winners. I held the cup up high over my head to show the crowd. I was embarrassed by how loudly and long my classmates applauded.

5

The Spark

After graduation, I applied for a job with a hot, new tech start-up in Cairo.

When my father first heard that I was going on an interview, he didn't hide his disdain: "A start-up? Rana, you graduated top of your class. You're better than this! Do you really want to go work at some unknown computer company?"

Then he uttered the words that made my heart sink: "I'm coming in with you."

Admittedly, Cairo is not as socially progressive as the West, but even in Cairo, having your father chaperone you to a job interview was bizarre. Eventually he relented: He waited for me outside in the car.

To my parents, there were only two legitimate career paths: working your way up in a major multinational, preferably one with a recognizable name and impressive headquarters; or earning a tenured position at a prestigious college or university. Start-ups were precarious. They couldn't guarantee a job through their next round of financing, much less for life. In a culture that is risk-adverse to its core, start-ups were viewed with suspicion.

But to my peers, the software start-up ITWorx was *the* place where everyone wanted to work. It offered a rare opportunity for educated, ambitious people in technology to remain in Egypt and feel a part of the changing technology landscape. There had been a major brain drain in Egypt and throughout much of the Middle East, in part due to lack of economic mobility. Many of my peers today work outside Egypt, for companies like Microsoft, Google, IBM, Cisco, and Intel, companies that were in the vanguard of the computer revolution.

In the mid-1990s, the Middle East lagged behind on the technology front. Microsoft was a billion-dollar company headed toward domination of the world's operating systems. The dot-com revolution was in full swing, and a Northern California valley renowned for making silicon chips had become a magnet for entrepreneurs and innovation. Very little of this was taking place in Cairo.

ITWorx was founded in 1994 by three entrepreneurs, Wael Amin, Youssri Helmy, and Ahmed Badr. Today, it employs more than eight hundred people throughout the Middle East and has a dazzling glass-façade headquarters. But when I was applying for a job, it occupied a modest space in a nondescript building in Heliopolis. True to start-up culture, ITWorx had a relaxed vibe: no suits or ties, just jeans and T-shirts. And as for office furniture, it didn't waste money that could be plowed back into the company: There were a few desks, a few chairs, and lots of people sitting on the floor talking and working.

I had dressed up for the interview in a very proper light-blue blouse with a Peter Pan collar, a blue floral skirt that hit below the knee, and low-heeled pumps; I even carried a pocketbook. I looked like a young MBA student. I was *anti*-hip.

After the initial introductions, my interviewer gestured toward the floor. I awkwardly slid down, carefully draping my skirt over my legs. For the next hour, I felt (and I'm sure *looked*) completely out of place, seated demurely with my legs folded beneath me, struggling to keep my skirt over my knees as we talked. I don't remember much about the job interview except I did not get an offer. Instead, I accepted a fellowship at AUC to pursue a master's degree, a decision that pleased my parents.

I did make an impression on one of the company's founders, Wael, who walked across the corridor shoeless, in socks, looked at me, and walked by. Apparently, he made a note to track me down and later asked my close friend, Hoda, who worked at ITWorx, to invite me to a barbecue at her home.

I was making some progress on the social front. I was now permitted to attend some well-chaperoned co-ed functions, and that's where I formally met Wael. Only twenty-two, he was wearing baggy jeans and a T-shirt. For someone who was shaking up the Cairo business scene, Wael was low-key, soft-spoken, and a bit of an introvert. But he was very approachable. He remembered seeing me at the job interview, and we spent a good deal of the party talking.

We bumped into each other again at an AUC employment fair, where ITWorx had taken a booth. I walked over to him and said, "Your booth is awesome, but I do have some constructive feedback." He laughed and replied, "Please email me your thoughts. People never give us feedback."

So, I did, writing a thoughtful and lengthy memo describing the five ways the ITWorx booth could be improved. And that was the start of our geeky romance. At the beginning, Wael would send me emails like "You know I'm reading an interesting book about the psychology of how to handle difficult people, have you read it?" And he would follow up by sending the book to me. And I would recommend books to him. I'm not sure which book recommendation captured his heart, but when I was on a family vacation in Alexandria later that summer, Wael called me on my very first cellphone.

I was in awe of Wael, a slightly older guy so smart and accomplished, who ran his own company and drove a cool BMW, and, to a green nineteen-year-old girl, seemed so mature. Wael had grown up in Argentina, where his father worked on a joint-government-sponsored project between Egypt and Argentina. His family moved back to Cairo when he was a teenager. Wael graduated from AUC at a younger age than even I did; I always regarded myself as smart, but I was convinced that Wael was even smarter, and definitely worldlier. I looked up to him. We not only clicked intellectually, but there was a real attraction between us.

After that, Wael and I started dating, in a modest Muslim fashion: in daylight and in secret.

On our first date—my first date ever—Wael treated me to ice cream, but the real magic didn't happen until our second date. Wael took me to lunch at the fancy Marriott Hotel in downtown Cairo with a stunning view of the Nile River. I didn't even notice the view. I sat across the table from Wael, our eyes locked on each other, and everything else seemed to fade away in the background. I was overwhelmed by a strange new sensation—I wasn't in love yet, that came later, but I was certainly infatuated.

This was all so new to me; it was exhilarating to finally allow myself to feel—and express—the emotions that I had kept under wraps throughout my teenage years. After all, I had already completed some important milestones in my life; I had my college degree and was working toward my master's, and I wasn't even twenty years old! I felt that I could cut myself some slack and have a social life.

First, like the nice Egyptian girl I was, I wanted to introduce him to my mother and sisters, but I swore them all to secrecy. (I knew that the minute my father got wind that I had a boyfriend, he'd expect to set a wedding date and hire the catering hall. Wael and I were not ready for that!) We met at Chili's Grill and Bar, an American chain restaurant known for its Tex-Mex cuisine, *the* hip place to go in Cairo at the time. Over heaping portions of Texas cheese fries, corn on the cob, and crispy chicken tacos, my sisters and mom pumped Wael for information about himself and his family.

He passed this first test—my mother and sisters approved of him.

Wael and I tried to steal time to be together whenever we could. Sometimes he would pick me up from AUC and we'd have lunch or an early dinner together, and then he'd drive me home. Or we would have coffee in a café near the college.

After we were dating for about a year, Wael, nice Egyptian boy that he was, proposed marriage . . . to my father. Although he was just twenty-three, Wael was an established and important member of the business community, but my dad didn't know what to make of him: "He owns his own company—a *start-up*. How stable is that? What if he runs out of money tomorrow? You'll be out on the street."

My dad also clued into the fact that Wael was quite the introvert and, like me, had very little experience with dating or romantic relationships. He was concerned that he might not be ready for marriage and needed to have some more life experiences before getting tied down. (Of course, that doesn't apply to a girl!)

My father refused to give his blessing until he knew more about Wael's family. First, he vetted the Amins with his friends; no one had anything bad to say. Then he hired a private investigator to thoroughly check out Wael's family. Wael's parents, Ahmed and Laila, came to our apartment at the villa in Heliopolis to meet my parents. The mood was formal, maybe even a bit frosty. I could see my parents sizing up the Amins, checking out how well they dressed, how well they spoke, and the social circles they ran in.

During their visit, the powder-blue landline rang. My father answered, and when he hung up, he had a smile on his face. The investigator he'd hired had given the match his seal of approval. The tension in the room instantly lifted, and we set a wedding date.

My life as an observer of human emotion, standing on the sidelines, was officially over: I had fallen head over heels in love, with the first man I dated or kissed. Wael turned my life around in more ways than one.

THE BOOK THAT CHANGED EVERYTHING

Even as I was planning my wedding, my parents encouraged me to continue my education. I was already enrolled in a master's degree program at AUC, but I couldn't figure out what to focus on. I didn't want to follow a conventional career path in computer science: I knew I wouldn't be happy spending my life tweaking operating systems, building faster processors, or designing videogames. In any case, I believed that there was a whole uncharted world to explore in terms of human-to-computer interaction, and I intuitively understood that computers could do so much more for us than we imagined at the time. I felt strongly that we had barely scratched the surface of their full potential, and I wanted to create something that would transform people's lives.

Wael understood my dilemma. He suggested that I read a new book that was getting a lot of attention in our circles, *Affective Computing*, by Rosalind Picard, an associate professor at the MIT Media Lab. I had never heard of the Media Lab, and knew very little about MIT, but I found a review of the book online and was fascinated. Computers should read and adapt to human emotions? How could silicon chips, glass, and wires understand what human beings were thinking or feeling? And why was this even necessary?

Picard was a respected engineer who, before joining the MIT Media Lab, had spent years working on artificial intelligence at Bell Labs. I wanted to learn more about her ideas, so I ordered the book.

At first glance, *Affective Computing* looked very much like the brainchild of an engineer, and one who was pretty conventional at that. But despite the subdued cover, the book was nothing short of radical, even revolutionary, reimagining the role of technology and its relationship to human beings.

"Emotions are important in human intelligence, rational decision making, social interaction, perception, memory, learning, creativity and more," Picard writes. "They are necessary for day-to-day functioning."

Just about every thought, action, and human interaction involves emotion. And, as Picard reasoned, if the point of AI was to design smarter computers that could emulate human thought and decision making, our machines would need more than pure logic. Like human beings, they would need a way to interpret and process emotion.

It didn't come as a surprise to me that people who were accomplished at "reading people" had higher EQs and were more successful in every aspect of life. After all, understanding the mindset and intention of another person is key to appropriate interaction. What I did find surprising, even astonishing, was the breadth and scope of the role of emotion in human life, which extended well beyond social interaction, spilling over into just about every human endeavor.

I was also struck by the vital role of emotion in enabling people to make sound decisions. At the time, I believed that the best decisions were based on cold, calculated logic that didn't let feelings get in the way. In fact, as I learned, decades of neuroscience showed just the opposite to be true. Your "feelings" don't get in the way; they improve your thought processes.

Sure, excess emotion (like road rage or paralyzing fear) can be harmful, but so can too little emotion. Perhaps the most compelling research in this regard was conducted by renowned neurosurgeon Antonio Damasio, a neuroscientist at the University of Southern California, who studied patients with sustained brain injuries that had disrupted communication between the left and right hemispheres. (The left side is associated with more linear, mathematical, logical reasoning—think IQ. The right side is associated with facial recognition, spatial identification, and more artistic endeavors—think EQ.)

For the most part, these patients maintained their cognitive intelligence. They could still read, crunch numbers, collect facts, and analyze problems, but they had no EQ. The injury that had blocked the connections between the "rational" brain and the "emotional" brain had rendered them unable to process emotion properly. And because they were unable to sort through their emotions in a meaningful way, they were severely handicapped in their ability to make any kind of decision, big or small. They struggled through life. They couldn't hold on to jobs. Their marriages broke up. They lost their money because they made terrible investments. They often ended up alone, broke, and unhappy.

As a computer scientist, I found it incomprehensible that we, knowing all we did about the essential role of emotion in brain function, were still modeling our mechanical brains on an antiquated view of intelligence. This serious fundamental flaw in computer design, I realized, would hinder both the potential of our machines and the human beings who were increasingly relying on them.

Picard's work left me inspired. We absolutely needed to build computers that could read and respond to emotions. I didn't have a clue yet as to how I would approach the problem, but I understood that this was important work. Moreover, I now had a focus for my thesis.

ALL ABOUT THE FACE

The waitress offered a weak smile (her zygomaticus muscles pulling upward, but her eyes did not crinkle) as her customer requested a detailed description of every item on the menu. The two men at the next table impatiently checked their watches. One shot her a dirty look (lip corner pulled up, eyebrows furrowed).

I felt bad for her. From a neighboring table, I was watching this mini drama play out as I sipped my tea and waited for Wael to join me. This was how I spent much of my time as a graduate student, studying faces.

Now that I had chosen my field, I saw people differently. When I looked at a face, I now saw beyond the lips, eyes, nose, and mouth. I saw an anatomical chart of the forty-three facial muscles that lie beneath the top layer of skin, the epidermis, deep down in the subcutaneous layer. If I concentrated on the face, I could see these muscles actually move, forming the lines, furrows, smiles, and frowns that create our facial expressions. And it felt as if I were seeing them in slow motion.

I was taking an online course to become a certified Facial Action Coding System (FACS) coder. Developed by psychologists Paul Ekman and Wallace V. Friesen in the 1970s, FACS is a system of mapping the facial muscles that lie below the skin to descriptors of those muscles. FACS is about the movement of muscles—how they lift, lower, twitch, and crinkle. But FACS doesn't link movements to specific emotions; it is all about the *mechanics* of facial movements.

In FACS, each facial muscle movement is called an "action unit." There are forty-six basic facial actions, spontaneous and subconscious shifts in facial expressions, like the rise of an eyebrow, the curl of a lip, or the wrinkling around an eye. For example, action unit 4 is the brow furrow (or pull of the corrugator muscle), the drawing together of the eyebrows. And action unit 12 is a lip corner pull (or pull of the zygomaticus muscle), which is the main component of a smile. If you smile and touch either side of your lips, you can feel the lips pull upward.

It takes about one hundred hours of training to achieve FACS certification. To pass the course, I had to learn all the action units and then put that knowledge to work. There was homework, too: I had to watch eighty videos of television interviews with politicians and actors, in slow motion, and code for every facial movement. As part of the training, you also practice making facial expressions in front of a mirror, which I did for hours at a time.

While FACS does not link facial expressions to mood, Ekman did use it to describe emotions. Through his research, he identified six basic universal emotions: anger, disgust, fear, happiness, sadness, and surprise (he later added contempt). He then correlated the action units to conform to these basic moods: AU12 + AU6 = happy. AU1 + AU4 + AU15 = sad, and so on.

The FACS muscle coding is not controversial; it is pretty much accepted science. But the concept that all facial expressions can be distilled down to six or even seven basic emotions is *very* controversial. When I first read about it, it bothered me, too. I didn't think it captured the full spectrum of our emotional or mental states. The beleaguered waitress I described earlier was displaying a polite but constrained smile, but she wasn't happy, or fearful, or contemptuous. She was somewhere between harried, annoyed, and distracted. My point is that there are more nuanced emotions beyond Ekman's six basic emotions—for example, distraction, curiosity, patience, confusion.

And then there are *degrees* of emotion. For example, if you tell me someone is angry, what does that mean? Is he furious, mildly annoyed, outraged? It needs clarification. Saying someone is angry is tantamount to saying, "I'm painting the wall blue." It prompts one to wonder, what shade of blue? Light blue? Royal blue? Navy blue? Cobalt blue?

Facial expressions also unfold over time. Think of them as phonemes or syllables that combine to create words and then sentences. To really understand the emotional and cognitive state of a person, you needed to map those expressions into emotion words or sentences.

I filed all my reservations in the back of my mind. The reality was that FACS was a real achievement in that it captured the ephemeral, it provided a language for communicating those fleeting facial muscle movements. For my purposes, if I were to teach a computer how to recognize human facial expressions (a window into our emotional state), I would need to break down every conceivable expression into its basic components. I would need a tool to convert smiles, smirks, and frowns into something a computer could process: quantitative data. The action units provided that system.

FACS training is hard: These forty-six facial action units can be fast and subtle, and can combine in a thousand different ways to portray hundreds of nuanced emotional states. Even human beings, who have an innate ability to interpret these signals, often miss significant pieces of the message.

When I broke ground on my master's thesis project, computers weren't just emotion blind, they were, well, *blind*. Today, you can snap a picture of your lunch, and the right app will identify what's on your plate, differentiate between a salad and a sandwich, and calculate the nutritional content with reasonable accuracy. This is the kind of highly sophisticated AI we could only dream about when I was a graduate student.

Back then, computer vision was primitive. Digital cameras were clumsy and slow; the images were gray and blurry. Remember those big, clunky webcams perched on top of your computer? That's what I had to work with. AI was still in its infancy. A top-of-the-line computer couldn't distinguish among a face, a frankfurter, and a piece of fruit.

Yes, I had a grand vision of inventing a computer algorithm that would be responsive to human emotions and needs, but the reality was that I was sitting in my lab staring at an emotionally ignorant piece of hardware. When it came to understanding human emotions, a newborn who knew how to find a "face" had greater skills than this mechanical brain. I would have to start by taking baby steps. If I were to get this "thing" to identify, quantify, and respond to human emotion, it would first have to understand a "face." So, for my master's thesis project, I decided to build a facial detector, that is, a tool that would enable a computer to distinguish a face from other objects. I spent a year building it.

The way to teach a computer to recognize a face—or to identify any object, animate or inanimate—is first to "show" it lots of faces, which means inputting lots of pictures of faces, all kinds of faces. And not just a whole face—I had to break the faces down into their components: the eyes, the brows, the forehead, the lips, and so on. Today, given the proliferation of images online, this is a piece of cake. But back then, there were very few images online, and hardly any faces. There was no Google Images, no YouTube, no Instagram, no Facebook. So, I tapped into one of the few data repositories of faces: the Cohn-Kanade Database, a collaborative effort between Jeff Cohn at the University of Pittsburgh and Takeo Kanade at Carnegie

Melon University. I emailed Dr. Cohn requesting permission to use their images, and he graciously allowed me to do so.

Using hundreds of students as models, the database offered pictures of people expressing the six basic emotions cited by Ekman. I downloaded this into my computer and wrote an algorithm to identify a face and also track it in a video. My master's thesis focused on building a face detector and a facial landmark detector: Think of it as a virtual face mask that identified the location of different feature points like your eyes, mouth, lips, and nose. My algorithm was predictive; it anticipated the new location of feature points, say the outward corners of your lips, based on their current acceleration magnitude and direction. My algorithm was particularly suited for subtle movements on the face as well as sudden head movements, both of which occur quite often in spontaneous expressions.

By the end of the year, I had completed my master's thesis. When I showed the computer a face—voilà! It placed a box around it. Back then, this was a really big deal.

Eventually, I would have to train the computer not just to locate the face, but to home in on the features (the eyes, the mouth, the brows), to track the movements, quantify them, and link them to facial expressions and then to mood. Then it would have to differentiate among subtle movements (a furrow of the brow, a slight curl of a lip). And then it would have to analyze, in real time, what that inferred, and then recognize the difference among a weak smile, a grin, and a contemptuous smirk.

But that's how science gets done: with the gradual, step-by-step progress you make every day and the inevitable setbacks you learn from. You don't have to be a genius to be a scientist, but you do need to be persistent.

6

A Married Woman

Wael and I were married on August 30, 2000, in a big wedding with six-hundred-plus guests at the Conrad hotel in downtown Cairo, on the banks of the Nile River. The reception was held in a giant ballroom with vaulted ceilings and crystal chandeliers, with a deejay, a singer, and of course the requisite belly dancer. There was no alcohol served, as is the custom in Islam, but it was a fun, high-spirited celebration.

The wedding was officially scheduled for eight P.M., but in true Egyptian style, I walked down the aisle with my father at eleven P.M. I wasn't a bit nervous about getting married. I was certain that I had found my *taw'un alruwh,* my soul mate. Rather, I was a Type-A Bridezilla, obsessing over whether the flower arrangements worked in the large ballroom, was the food being served properly, was the band good enough. I guess there's a part of me that is very much my father's daughter. When it came down to it, I was fretting over "What will the guests think?"

Wael and I danced to Shania Twain's "From This Moment" and

spent the rest of the night on the dance floor before heading off to a seaside resort in Bali for our honeymoon. A month later, we returned to our Cairo apartment as a married couple. The marriage was everything I had hoped it would be. We were lovers but also thought partners, linked together by common beliefs and goals. Wael was my adviser and career mentor, and just as we did during our courtship, we discussed books and the latest technology and planned for our future. I trusted Wael's judgment, and pretty much did what he told me to.

As a modern, progressive two-career Muslim couple, we had few role models to follow. I knew that I did not want to duplicate my parents' marriage. I wasn't going to wait on Wael the way my mother waited on my father and, in truth, the way Wael's mother waited on his father. We had to invent our own version of marriage, carve out something new and different. We were both inexperienced on the relationship front. I had never dated before marriage, and while Wael had dated a bit, he had never been in a serious relationship. We were trailblazers, figuring out the new roles of men and women in a new Middle East.

Wael and I were both impressed by Stephen Covey's 1989 book, *The 7 Habits of Highly Effective People*, in which Covey recommended that individuals should write personal mission statements modeled after the organizational mission statement. So we came up with the idea of creating a mission statement for our marriage. This may sound a bit nerdy, but in fact, for us it made sense. We were both highly analytical, and making lists came naturally to us. We sat down and wrote out our goals and aspirations for our marriage.

We wrote, "We want to be the hub for our community: our family, our friends, and our co-workers, and be a force of positive impact." We made the decision to hold off having children because we loved to travel and we wanted to have fun for a few years. This would also enable me to continue my education if I wanted to pursue a PhD.

We strived to be a positive influence for the people around us. We saw our home as a place where people would congregate for potluck dinners and exchange ideas; we wanted to be open-minded and preserve the best of Muslim teaching, but also embrace the modern ways of doing things, too.

Despite all the bullet points and corporate lingo, we were very affectionate. We would even kiss in public, which was highly unusual for a Muslim couple. Once, at a dinner party at a friend's house, one of my friends asked me why I never wore lipstick. (I don't wear makeup at all.) Another friend commented, "Why should she bother? Her husband would just kiss it off!" Everyone laughed. I felt loved and cared for, and for a long time, Wael and I were considered the perfect couple.

We lived next door to my in-laws, and on Fridays, the Muslim Sabbath, after Wael and his father went to pray, we'd have lunch at their home. I grew very close to my in-laws. I called them Tant Laila and Uncle Ahmed, and developed genuine affection for them, and they for me. They treated me like a daughter; I wanted to please them, which may partially explain why I took the next big step in my life.

THE ROLE OF FAITH IN EGYPT

The millennium ushered in a new spirit in Egypt, especially among the young, that was both traditional and conservative. A religious revival swept through Egypt, led by a charismatic Muslim activist and TV preacher named Amr Khaled, whose message resonated deeply with young people throughout the Middle East. For nearly a decade, his was one of the most popular TV shows in the Middle East.

Khaled was nothing like the other evangelical imams, typically grim graybeards who dressed in religious robes and threatened eternal damnation. He was handsome and soft-spoken, with a trim mustache, sharp Western suits, and an upbeat, optimistic way of speaking. He spoke lovingly about his wife and the need for men to be kind and understanding toward women.

As modern in outlook as he was, Khaled held more conservative views on modesty for women. He urged all women to wear the hijab, but he made it sound more like a prize than a burden. "The most precious thing a woman possesses is her modesty. And the most precious thing of modesty is the hijab. If I asked you a question, if I asked you what is the most precious thing you owned, what would that be? If you have something precious, will you take care of it and protect it?"

Many women saw the hijab in a new light, as an affirmation of their own self-worth, a symbol of their value, as opposed to something keeping them in their place. It made a big impression on many young women, including me. Even some of my modern-thinking friends from college were now proudly wearing a hijab.

What I liked most about Amr Khaled's teachings was his emphasis on the core values of Islam: hard work, respect, love, and honesty.

I asked Wael how he felt about me "taking the veil." After all, when he married me, I was not a hijabi woman. But he said that it was entirely my decision.

On December 1, 2000, for the first time in my life, I wore a hijab out in public and accepted the code of modesty. I wore a lovely orange and olive floral scarf with olive slacks, the epitome of Muslim chic. Typically, the first time a woman appears in public in a hijab, friends and family congratulate her, almost as if she has reached a significant milestone in her life. I got a lot of positive reinforcement, except from my project partner at AUC where I was teaching, a modern-thinking woman. When I walked into her office wearing a hijab, she was shocked—maybe even horrified. And she challenged my decision. But six months later, she, too, was wearing a hijab.

For the next twelve years, I was never seen in public without a head covering. I covered my arms, too, and did not wear a bathing suit in a co-ed setting. Several years later, my mother and my two sisters also took the veil.

The fact that so many fashion-conscious young women were now wearing a hijab ushered in a reimagining of the headscarf: Women opted for beautiful patterns and fabrics, matching them with fashionable outfits. This was not my grandmother's hijab, or my aunt's drab niqab.

The religious fervor continued for about a decade. As often happens, the pendulum has swung back in the other direction. Khaled's star has faded. Many of the women I knew who had followed his teachings have long since abandoned their hijabs, as have I. Once frowned upon, dancing is back in style at weddings. It took a dose of true religious conservatism under the brief reign of the Muslim Brotherhood to sour women on it. Once the way one practiced religion became an edict instead of choice, it quickly lost its appeal.

I know that science and religion are supposed to be polar opposites, but I don't see it that way. In science and religion, "believers" have to imagine the world not as it is, but as it will be. Of course, at some point in science, you do have to produce proof of concept, or risk losing funding, your academic standing, and your reputation. But at the beginning of your work, it is your belief, your faith, that propels you forward.

When I began my master's thesis, I was a true believer, or I never could have pursued the project the way I did. I had a vision of smarter technology that could better serve humankind, that understood human beings and our emotional states, allowing us to interact naturally and effortlessly. Much of the technology I needed in order to fully achieve this vision had not been invented yet. I was taking a tremendous risk, and that required faith.

LEAVING HOME

I had overcome one roadblock to building a computer that recognized emotion—my algorithm could now identify and home in on a human face—but I still had a long road ahead to achieve my goal of creating an *emotionally intelligent* computer. It would have to recognize facial expressions, quantify them, link them to the correct emotional state, and then respond appropriately.

For me, the next logical step was to pursue a PhD in computer science so I could continue my work. My goal was to eventually join the faculty of AUC.

AUC had not yet hired any tenure-track female computer scientists. In fact, getting a tenured position in any department was a major challenge. If I were to break that glass ceiling, I would need to follow the path of the men who got tenure-track positions. Most had earned PhDs from top universities in the United States and Europe. I was first in my class and had gained a reputation as a talented scientist. Still, I was a married woman. To achieve my goal meant leaving Egypt, but that was complicated. Wael couldn't come with me: He had a company to run in Cairo.

If I had been a man, there would have been no question that my spouse would follow me wherever I had to go. But in my culture, I was on very shaky ground. Married women do not move halfway around the world to pursue their dreams. Such a thing is not accepted or understood. I was positively gobsmacked when a female CEO I met in the States told me casually over coffee that her husband stayed home with the kids. In Egypt, that would be inconceivable.

Fortunately, Wael was forward-thinking enough not to hold me back. In fact, he encouraged me to apply to schools overseas to assess my prospects. U.S. schools were out of the question because they were too far away for frequent visits. So, I applied, very late in the game, to the most prestigious schools in the United Kingdom with the best computer science departments. This wasn't just a matter of filling out forms online. I had to write a complete proposal describing what I wanted to design and build, and a precise blueprint for how I was going to do it.

My first choice was the computer laboratory at the University of Cambridge, the school where mathematician Charles Babbage conceived of the modern computer more than two centuries ago. I applied for a fellowship with Peter Robinson, who headed the Rainbow Group at the computer lab. The Rainbow Group was on a mission to reimagine the interaction between human beings and computers, a mission that dovetailed nicely with my concept of building an emotionally intelligent computer system. Those who work in human-computer interfaces (HCI) design the keyboards, touch screens, and now voice-activated computers that enable the everyday individual with no technical background to engage easily with a computer.

Because I had applied so late, I honestly didn't expect to hear back. I thought I would be asked to reapply for the following year. Then, in early August, I received an email from Professor Robinson, congratulating me on being accepted to the PhD program at the Cambridge lab. Not only had I been accepted to the top computer research center in the world, but I was offered a full scholarship.

I should have been ecstatic. Instead, I panicked. With the beginning of the new academic term less than a month away, I didn't have a lot of time to mull the invitation over. I had just a few days to accept the fellowship or decline it.

Part of me desperately wanted to stay home in Cairo with my husband and continue our blissful, honeymoon-like existence. But another part of me understood the magnitude of this opportunity, and where it could potentially take me. Unable to make a decision, I called Dr. Robinson and explained that although I was thrilled to have been accepted into the program, there was a complication: I was married. Would he allow me to work remotely from Cairo? I assured him that I would visit the lab regularly, but preferred not to move to Cambridge.

Peter was sympathetic to my situation, but his answer was an emphatic no. He expected me to do my work at the lab, like all the other doctoral candidates. Looking back, I see that he was absolutely right. I could not have done what I did working remotely on my own. And I would have missed out on some pivotal relationships that shaped my research and my work.

Wael believed that it was the chance of a lifetime, and I was leaning toward accepting the offer, but I needed some reassurance that I was doing the right thing. I went to talk my decision over with my parents. Given the way they had always prioritized our education, I thought they would encourage me to get my PhD from Cambridge. But that afternoon, I learned the facts of life.

It's true that my mother worked while raising a family, and was respected in her profession as a trailblazer. To me, she appeared to have everything, and to have achieved parity with my father. But appearances can be deceptive.

My father revealed to me the reality of my mother's life, one that I had been too blind to see. Even though she had enjoyed a successful career, her work was secondary to everything else she did as a wife and mother. When she walked through the door to our home, she left her business persona behind. She never talked about her job, or took work calls at home. As smart and capable as she was, she couldn't have "leaned in," because my father's expectation was that at three P.M., when school let out, she would be home taking care of her daughters.

I learned that my mother was never "allowed" to go on a business trip by herself. If she was asked by her boss to visit a client overseas or attend a training seminar in England, she would decline the opportunity unless the trip fell during a school holiday, when our family could accompany her and turn it into a working vacation. How had I missed this? I remembered now that from time to time, when we were on one of these "vacations," my mother would leave us for a few hours to attend a training program while my father took us sightseeing. It had never dawned on me why we were all there. Nor had I questioned why my mother never mentioned her work at home.

While talking to my father, I recalled an incident that occurred when I was around six. My mother had come home from work with a pager that had been given to her by her boss, who wanted to be able to contact her after hours in case there was a problem at the bank. It was a sign that her role was highly valued by her boss and that she was poised for promotion. My father, however, was furious. He told her she couldn't keep the pager. So the next day she brought it back to the bank, and her boss gave it to someone else. At the time, I couldn't understand why my father was making such a fuss over what I thought was a cool piece of technology; after all, he loved this stuff! Now it all made sense. My mother's success in her career was fine as long as it did not interfere with family life or inconvenience my father in any way. The pager meant that someone else had claim over my mother's time and that was simply unacceptable. I wonder how my mother must have felt having to pass on the pager and those other opportunities. But she never complained. She buried her ambition and accepted her secondary role as a way of life.

Was that to be my destiny, too?

"Rana, you're a married woman now. This decision is between you and Wael. But you know where we stand. We don't think you should go."

Translation: "You're under Wael's jurisdiction now, not mine."

For a nice Egyptian girl who never disagreed with her parents on anything important, the thought that I might do something to disappoint them caused me anguish. But my father had acknowledged that this was between Wael and me.

If Wael had no objections, I decided I had to follow my heart. *Rana, if you don't do this, you'll regret it your whole life.*

I prayed and cried over it, but eventually, I accepted the fellowship. I felt grateful that I was married to a man like Wael, who was open-minded enough to allow me to pursue my dream.

I was scheduled to fly to London on Tuesday, September 18, 2001.

Late in the afternoon on September 11, Wael and I were in our living room when we saw a news flash on TV. The Twin Towers at the World Trade Center in New York had crumbled to the ground after being hit by two commercial airplanes, and the Pentagon in Washington had been attacked and was on fire. As the tragedy unfolded, we were riveted to the TV. The events seemed close enough to be in our backyard. I knew intuitively that when something of this magnitude happens, it impacts the entire world.

Then my phone started to ring.

"You can't go to Cambridge now," my mother declared. "We're on the brink of a third world war. You'll be a Muslim woman stuck in the West. You'll be a target."

Everyone in my family advised me to cancel my plans. People in Cairo believed that this would blow up into a world war, West versus East. But once again, Wael, who had traveled throughout the States and understood the culture better than my family, was adamant that there wasn't going to be a world war. And if I didn't seize this opportunity to get my PhD from Cambridge, I would never do it. And without a PhD, I would never fulfill my dreams.

I was terrified of traveling to Cambridge, given the political climate. I had no idea of the kind of reception I would get.

My family threw a farewell "party" for me at our apartment the weekend before I left. It was a grim affair, with more tears than smiles. No one hid their concern over my safety and my judgment. My mother-in-law, her expression pained, pulled me aside and carefully placed a necklace around my neck. It was decorated with Quranic inscriptions, to protect me from harm. It was a bit unsettling to see how afraid she was for me.

A close friend warned me that going to Cambridge would wreck my marriage. Others worried that I would be too lonely to function. But I dismissed their concerns. I felt so confident in my relationship—and I think Wael did, too. We didn't believe that time or distance could tear us apart.

PART II

The Scientist and the Mind Reader

7

Stranger in a Strange Land

As I sat in the cab to the airport with Wael that morning, to catch a flight to London, my emotions were bouncing all over the place. I was elated at the prospect of embarking on a new adventure, bringing my vision of an emotion-smart computer to fruition. But I was also frightened by the uncertainty of what lay ahead. What would it be like living in Cambridge on my own? Would I be singled out as a Muslim? But mostly, I was in a state of disbelief: *Am I really going through with this?* Had I, as a nice, well-mannered Egyptian girl, crossed the line?

I was determined to forge ahead, but not without some misgivings. Wael and I would scarcely see each other for the next three years. He could stay in Cambridge for only a couple of days, to help me get settled. Just the thought of his leaving triggered a wave of homesickness in me.

Maybe it was the protective amulet I wore around my neck, but after all the angst, all the warnings from family and friends of the potential dangers facing a Muslim woman in the United Kingdom, my journey was anticlimactic. After the five-and-half-hour flight, the drive to Cambridge took place mostly on nondescript highways, the scenery gray and industrial. We were let off in central Cambridge, and Wael and I walked down a quaint cobblestone road, carrying my two suitcases to a hotel on the River Cam. We were hungry and tired. We stumbled onto a hole-in-the-wall Japanese noodle bar, which quickly became our place when Wael was in town.

Cambridge is one of two universities in the world—Oxford is the other—that operate under a collegiate system, meaning its eighteen thousand students are divided among thirty-one colleges. Although the university is very large, each college feels distinct and personal, and serves its own community. The college is your home base, where you live, dine, and hang out between classes.

Newly accepted students must choose to affiliate with one of the thirty-one colleges. I was completely naïve about the system. A savvier student might have leapt at the opportunity to affiliate with one of the older, more prestigious colleges, like King's College or Trinity (whose alumni include Sir Isaac Newton) or St. John's. I selected my college based on the only criterion that mattered to me at the time: It was all female.

Nearly all the colleges were co-ed, which meant that male and female students lived together on the same premises. Given the way I had been raised, I didn't approve of young men and women living next door to each other in an intimate setting like a dorm. In the Middle East, Muslim men and women who are unrelated and unmarried adhere to a strict code of conduct in terms of how they interact with one another. Living side by side in close quarters is not permissible; neither is casual touching.

At the time, I would never have hugged or even planted a friendly kiss on the cheek of a man who was not a relative. In the West, I discovered that everyone hugs and kisses one another all the time. I remember the first time a male colleague gave me a friendly hug hello. It was just a casual sort of half hug, one that he gave everyone. Although I didn't let on at the time, I was so shocked by his action that by the time I got back to my apartment, I was shaking. When I closed the door, I burst into tears. I was still struggling to reconcile being a nice, *religious* Egyptian girl while fitting in and being independent.

So, that's how I ended up at Newnham College, one of two all-women colleges at Cambridge. As I was to learn, Newnham College had a rich history. It had been at the forefront of the movement to make higher education available to women, and it turned out to be a good fit for me.

When Cambridge was founded in 1209 by two ex–Oxford students, it was, like every other important institution of the day, all male. It took another six and a half centuries for the school to open Girton, its first college for women, in 1869. Two years later, Newnham College was established as a safe residence for "intellectually curious" women who wanted to attend lectures at Cambridge— "Lectures for Ladies," as they called them. But Cambridge didn't offer women degrees at either college until 1948. I am embarrassed to say that I was so sheltered that I had never heard of any of Newnham's illustrious alumni, including Rosalind Franklin, the chemist and crystallographer who provided essential information for the decoding of DNA, although the Nobel Prize awarded for the accomplishment in 1962 went to three men: James Watson, Francis Crick, and Maurice Wilkins. Typical! A residence hall for graduate students now bears her name. Other notables include Jane Goodall and Sylvia Plath.

Newnham is farther out from the Cambridge city center than most other colleges. The campus is a breath of fresh air, with expansive, spectacular green lawns (some seventeen acres) that students can walk on and beautiful, well-tended English gardens that spring to life in April.

Newnham was everything I imagined a proper English college should look like: built in the Queen Anne style (redbrick buildings with gables, white windows, pointy roofs, and bay windows). But what really distinguishes the campus, as the Newnham website describes it, is its "series of elegant halls linked by corridors and set around the College's gardens . . . the second longest continuous corridor in Europe," which enables students to navigate among the main buildings without stepping outdoors. Considering just how raw Cambridge winters can be, it would have been a big selling point for the school. (A school that's good for your brain *and* your hair!)

When Wael and I first arrived on campus, I was struck by its quiet beauty. I had been assigned a private room in an all-women dormitory for graduate students, as I'd requested. As I was about to sign my housing agreement with the college, I mentioned to the administrator in charge of housing that I was married. My husband would be visiting from time to time, staying overnight, I said.

The housing tsar gave me a stern look and shook her head. "That's just not possible. We have a strict curfew here for men, married or not. Your husband has to be out of the residence by eleven P.M."

A curfew for husbands! That's one even my father would think extreme. Clearly, I couldn't live in the dormitory unless I moved to a hotel whenever Wael visited. And given that Wael wasn't going to live with me full-time, I wasn't eligible for married student housing. All this meant that I had to find another place to live, quickly. This was before smartphones and tablets, so we couldn't ask Siri or Cortana for a list of available studio apartments in the area.

Back then, you could not access the Internet without logging on to an old-school computer with a modem. There weren't yet hotspots where you could sign on with your laptop. Instead, there were Internet cafés, loftlike spaces with row upon row of computers. Fees were based on the amount of time you spent surfing the Net. (Internet cafés still exist in some parts of the world.)

Near the campus, we found an Internet café crowded with students and grabbed a seat at an open computer in the back. Google barely existed then; we used Yahoo to search. When I typed in "Cambridge studios for rent," the top searches kept spitting back listings for studios in Cambridge, *Massachusetts*. I was becoming increasingly annoyed.

"Doesn't Yahoo know that the original Cambridge is here, in the UK?" I asked Wael indignantly.

"Actually, no," he said. "It has no idea where you are."

And, of course, because the computer I was using was emotion blind, it was clueless to how frustrated I was getting, which might otherwise have been a tip-off to it that it was offering me the wrong information.

We finally located a second-floor studio apartment that was a twenty-five-minute bike ride away from the Cambridge Computer Lab. We contacted the landlord, and raced over to see it. It was an unfurnished efficiency, one big room with a kitchen, living room/bedroom, and bath—fine for my purposes. So, we signed a lease and furnished the apartment in one afternoon using an inexpensive furniture store in Cambridge.

I experienced a deep sense of loss when Wael left to go back to Cairo. It finally sunk in that our honeymoon year was over, and for the next three years or so, we would see each other at most for a week or two at a time. As I sat in my Cambridge apartment alone that night, a thought occurred to me that struck fear in my heart. In Islam, we believe that Allah will test every one of us—believers and nonbelievers alike. The first category of tests revolves around your degree of believing and obeying Islamic rules. But the second category of tests is even harder: "Verily, We shall put you to test with some fear, and hunger, and with some loss of wealth, lives, and offspring" (2:155).

This is a test of personal faith; despite these losses, will our faith remain unshaken?

The hardest test Allah could inflict on me would be taking Wael away from me. I prayed for his safety. And I prayed that we would survive being apart and that it wouldn't hurt our relationship. I recited a prayer from the Quran, which roughly translates to "Allah, please do not test me on that which I cannot handle; I ask for your forgiveness and compassion."

The next morning, however, I didn't allow myself to sit around and mope. I switched into work mode. I refused to feel sorry for myself or wallow in my loneliness. That is not how I was raised. I had a job to do, and I was determined to give it my all. Still, it took a while to settle into my new life.

Overnight, my world went from mango-colored walls, blue skies, and brilliant sunshine to a bland carpet, beige walls, and a beige-y tan couch. I had no TV, radio, or CD player. When I walked out the door, most mornings I was greeted with a damp, gunmetal-gray sky.

The sleek new computer lab was designed for efficiency, with clean, open spaces, bare walls, good lighting, and lots of metal and glass. My office was also unadorned: a desk and chair, the surfaces clear of decorations. I did not even have a picture of Wael or my family on my desk. The message I had created in my workspace was clear: I was in work mode here, with no distractions or personal touches.

Living on my own, I learned every day just how sheltered I had been previously. When I lived with my grandparents, there was always somebody else around to clean up and do my laundry. When I was married, Wael and I had a housekeeper who came every day to clean our apartment. Help in Egypt was cheap, and this was not unusual for upper-middle-class citizens, even newlyweds. When I signed the lease for my apartment in Cambridge, my first question to the landlord was "Who cleans the apartment?" When he responded, "You do!" I was floored.

So, I rolled up my sleeves and learned to scrub down my own kitchen. Twice a week, after work, I would lug my laundry bag to the closest Laundromat, a couple of blocks away. I would sit in the Laundromat reading—usually a textbook about the science of facial expressions and emotions—while the laundry cycle was completed. This trek took some getting used to, and it got progressively harder as the weather became colder and windier. It was a far cry from my life in Cairo, where so many mundane tasks like this were all magically taken care of for me.

Life was harder and harsher than it was in Egypt; I didn't have my support network: my mom, my sisters, my in-laws, my friends. Yet, I embraced my new circumstances as a challenge. I was being given a crash course on how to become independent and figure things out on my own.

I didn't want to attract attention to myself by wearing my hijab around Cambridge or the lab, but I still wanted to maintain my modesty by keeping my head covered when I was indoors and around men who were not relatives. So, I decided to get around this by replacing my hijab with a hat. I bought a few beige and brown fedora-style hats, but the one I wore the most was a purple-and-pink striped fedora with a small brim that sort of covered my forehead. A strange choice, I guess, for someone not wanting to call attention to herself. I must have looked very odd wearing it indoors, but the uber-polite Brits never commented on it. Three months into my fellowship, I felt secure enough to put the hijab back on, and no one ever said anything about that, either.

I lived across the street from the River Cam, which ran through Cambridge. Cambridge was a far more rural town than I had imagined. To a city girl used to subways and concrete, this was the country. On nice days, I walked along the riverbank, watching the rowers glide by, enjoying the tranquility and fresh air.

One evening, months later, while I was strolling along the riverbank, it started to snow. I had seen snow in movies and had read about it, but I had no idea what it would feel like. I was enchanted by it. I absolutely hated the cold, but I lingered outside until I was covered in white.

Although we had exchanged numerous emails and even talked by phone, I had not yet met my thesis adviser, Peter Robinson, a tenured professor and vice head of the Cambridge Computer Lab. Horror stories abound about abusive thesis advisers who virtually enslave their doctoral students, steal credit for their work, and do everything to keep them from moving on. So, I was a nervous wreck at the prospect of meeting him face-to-face.

As I left my apartment for my first day at the Computer Science Lab, I had the usual first-day butterflies. I dressed in my lab uniform: beige slacks, a blue top, and "the hat." It was a bright, sunny day, and I rode my new bicycle to the lab, about a mile away from my apartment.

When I arrived on campus, as I was pulling up to the bicycle shed to store my bike, a man in khakis and a blue sweater pulled his bike up next to mine. He had a much older, dowdier bike, with a quaint wicker basket in the front brimming with books. He extended his hand and said, "Hi, Rana. I'm Peter."

I was speechless. One of the foremost computer scientists in the world was humble and friendly. And most shocking of all to a girl from the Middle East, he wanted me to address him as "Peter." In Egypt, a man of his stature would be driving a Mercedes and would insist on the best parking spot on campus, and I would have been expected to be deferential to him. None of that sense of privilege existed at Cambridge, at least not among my colleagues. Peter, I discovered, treated everyone as an equal. It was one cultural difference I appreciated.

My initial impression of Peter proved to be correct. Over the years, I found him to be supportive, modest, and generous about giving credit to his students; he was also, without fail, his PhD candidates' staunchest cheerleader.

Although I still saw myself as the nice Egyptian girl seeking everyone's approval, for the first time, I was confronting real skepticism about my project. Particularly daunting was the fact that some very intelligent, experienced computer scientists were telling me that the current technology was not nearly sophisticated enough for me to pull it off. If that proved to be the case, my career would be short-lived.

There was some truth to what they were saying. Not much had

changed since my master's thesis: Digital cameras at the time were big, boxy, and slow. Computer vision was just getting off the ground. And the AI tools that proved to be critical to my work (machine learning and deep learning) were still pretty clunky. Several peers warned me that I was wasting my time, that I should choose another thesis, one that had a better chance of being completed in three years. Translation: You may not end up with a PhD.

The thought of failure concerned me, but it didn't stop me. This project wasn't just a means to an end, to earning a PhD. This was something I deeply believed in, enough to turn my life upside down to pursue it. Scientists breaking new ground often encounter headwinds; you can't let them stop you, but you can't ignore them either.

Given the complexity of my project, I realized that I would have to be highly strategic in how I approached my work. Building a face reader—or, as I called it, a mind reader—would require input from many different experts in the computer lab. And I needed to line up that support. But the criticism forced me to develop a thicker skin, and sharpen the way I talked about it, two life skills that proved to be essential down the road.

Among the people at Cambridge whom I needed to have on my side was Sir David MacKay, head of the Cavendish Lab and inventor of Dasher, a keyboard-free tracking system for computers that enabled people with severe disabilities (like cerebral palsy) to interface with a computer using eye movements, head movements, or even breathing. This research was especially pertinent to my project because head gestures and eye movements are critical components in the expression of emotion. Another renowned scientist, Andrew Blake, a computer vision and machine learning guru and laboratory director of Microsoft Research at Cambridge, later became a co-adviser on my project. Dr. Blake was head of the team at Microsoft that built the Xbox. Computer vision (the ability of a computer not only to "see" a face but to home in on individual features and the subtle shifts in facial expressions) is fundamental to my work.

To his credit, Peter was curious about my project—no one else at the lab was working on anything like it—but he wasn't exactly passionate about it. Still, he was open-minded, which was why I could continue my work.

I realized I wasn't yet conveying the story of emotional intelligence in computers in a way that resonated with people. If many of my peers were questioning my choice of project, it was incumbent upon me to do a better job of explaining why I'd chosen it, and why I was willing to invest three years in building it. I had to double down on my efforts to persuade others to break through those barriers.

IT, ME, AND WAEL

Wael and I were used to communicating through a screen. While we were engaged, we would exchange emails every night, or use an old-school form of texting. We may well have had the world's first cyber romance. While other couples celebrated the anniversary of their first meeting or their first kiss, Wael and I celebrated our thousandth email exchange, a milestone we reached while still dating. (He surprised me with a romantic sunset cruise on the Nile.) Of course, in between all those emails and text messages, we had frequent face-to-face contact. But I thought that somehow, even when I was in England, we would be able to share our feelings and experiences via the Internet.

Indeed, my whole life was a series of computer screens, from my laptop at home to my desktop in my office at the lab. At lunch, I would grab a plate and sit down with the other PhD students, but in the evenings, I was usually one of the last people to leave the lab. As one of the few women at the lab, working on a project that many considered futile, I felt enormous pressure to prove myself.

Several weeks after Wael had left, it began to sink in that I would be all alone for the next three years. I was eating dinner in front of my laptop, doing research and working on my notes, feeling desolate.

Long-distance calls were expensive, so Wael and I would instant-message each other at night, via ICQ Chat, a window that popped up on the side of the screen.

Wael would check in when he finished work:

WAEL: How are you doing?
ME: (typing) I'm fine.

Really, I was miserable.

WAEL: How's the work going?
ME: Oh great, I'm reviewing right now, and getting ready to present in front of my group next week. Can't wait! How did your day go?

Actually, I didn't think anyone at the lab appreciated my project. I had no idea what to say in my presentation.

WAEL: I had some meetings that went pretty well. I guess you're busy; I'll let you go. Let's talk tomorrow.
ME: OK, sounds good.

As I typed my last message, I started to cry. I could see my reflection on the screen, the tears streaming down my face. Why couldn't he sense that? Texting my feelings felt phony somehow, inauthentic. I was looking to Wael for support, but how could he give me the kind of support I needed if he had no idea what I was feeling?

The computer was my main portal of communication with him, but my raw emotions and the second-by-second shifts in my mood were disappearing in cyberspace. He couldn't see what was going on in my head. The silent conversations I had with myself were invisible to him. Nor did I have any way of knowing what *he* was thinking.

Scientists talk about an "aha moment," a "jolt of discovery" when the pieces to a complicated puzzle somehow fall together. For me, this was one of them. I realized that computers were rapidly shifting how we communicated. We still turn to them as handy tools to solve problems (finding an apartment, checking the weather, or typing out a quick message to a colleague), but they were also fast becoming the major conduit through which we communicated. The challenge was no longer about human–computer interface; it was about human–human relationships. The computer was quickly becoming the mediator of our interactions. And as my experience was showing me, the emotion-blind computer was failing miserably at the task.

Although my computer and I were inseparable on one level, there was a chasm between us. It had no idea who I was; I could be any user who logged on. Nor did it know anything about what I was feeling, or what made me tick. It didn't know whether I was having a good day or a bad one, whether I was smiling or frowning, happy or angry, interested in what I was doing or bored out of my mind. Over and over, it spat back data in the same brittle, impersonal, generic way, regardless of my mood or the mood of the person I was communicating with.

All those emails that were sent between Wael and me were just empty words on a screen. My husband had no idea how much I missed him, or how hard our separation was for me. Nor could I tell how he felt about being left alone in Cairo—if he was happy that I was pursuing my dream or, beneath the surface, resented my absences. The computer that could spit out messages 24/7 had created the *illusion* of connection. But as I would come to understand, there is no true connection without the sharing of emotion.

8

A Mad Scientist Talks to a Wall

I was in the awkward stage of the innovation cycle, the incubation period, aka "the Mad Scientist Phase." In my mind's eye, I was practically holding the Mind Reader 3.0 in my hands, a machine so attuned to human emotional states that, on this October morning, the day of my first formal presentation in front of the entire lab, it would be supportive of me.

Hey, Rana. I see that you're feeling a bit anxious. Let's run through your talk again.

A machine that was so intuitive that when I walked into my office, it would dim the lights without missing a beat and turn on soothing "spa music" while it guided me through five minutes of mindfulness exercises to help me focus. And my empathetic social robot would hand me a cup of black *shai* (Egyptian tea).

Okay, Rana, back to reality.

In fact, Mind Reader version 1.0 was still on the drawing board. When it came to emotional intelligence, computers were as socially savvy as doorknobs, and the dream was still just that: a vision of what might be that only I could see.

I knew where I needed to land in three years, but there were significant hurdles yet to be overcome. If all went well, first the dream would become a prototype, elevated from an amorphous concept to something concrete. But at that point, the real work would be just beginning. Now the inventor would have to roll up her sleeves and begin tweaking, improving, and revising the software. And once I'd finished my first iteration, I could see what I'd done wrong and start all over again (and again), improving it. Depending on the complexity of the project, it could take several years before it was ready for prime time.

It was obvious to me that, given the expanding role of computers in our lives, equipping them with EQ was an absolute necessity, but many of my colleagues believed just the opposite: that the lack of emotion, the "clear-eyed," calculated objectivity of a computer, was what gave these machines their edge over human beings. So, I understood that for my first official introduction to the group, I had to make a strong case for humanity.

Steve Jobs is often quoted as saying, "People don't know what they want until you show it to them." I'm sure that worked well once Jobs had a sleek, skinny iPhone to wave around. (Who wouldn't want an iPhone over a stodgy flip phone or the pudgy smartphone competitors of the day?) But given that I was still "incubating" my Mind Reader, most of the working knowledge was in my head. Without a cool working prototype to show off, I had to dream up another way to capture the imagination of my audience and demonstrate why my project was worthy of a PhD from *the* Cambridge Computer Lab, that it was not only groundbreaking but also absolutely essential.

A few nights earlier, while "communicating" with Wael in real time, exchanging text messages in rapid succession, I had never felt more cut off from him or from my own feelings. I'd never felt more alone. For the first time, it really hit home how emotionally adrift I was when attempting to communicate in cyberspace. I was as emotionally blind, as emotionally cut off, as my computer. My goal in my presentation, then, was to make every person in that auditorium feel as disconnected as I did while trying to carry on a meaningful conversation without the benefit of face-to-face interaction.

Most techies are assumed to be introverts, but that's not me. I love speaking in public; I come alive in the spotlight. So, I was both excited and nervous at the prospect of meeting the people with whom I would be spending much of my time over the next three years. And like the nice Egyptian girl I am, I wanted them not only to like me, but also to respect my work.

Some people can glide into their speeches and presentations cold, without notes or preparation, and wing it. I approach each public talk the way I approach my work: I choreograph it carefully and methodically. I prepare, rehearse, refine, review, again and again, in the shower, at breakfast, on the bus on my way to work. Indeed, I am prepping (iterating, tweaking) right up to the moment I begin the presentation. Today was no exception. I would be speaking in front of some of the top computer scientists in the world, and my presentation needed to be pitch perfect.

We were meeting in a small auditorium at the lab, and I arrived early so that I could go over my presentation a final time. I entered from the front of the auditorium and waited for my colleagues to file in and take their seats. The crowd of fifty or so was overwhelmingly male. There were perhaps two women in the audience. Peter, my adviser, was in the center seat of the front row.

When everyone had settled down, I stood up, looked over the crowd for a few seconds, and then, just as I had plotted out days earlier, I slowly turned around so that my back was facing the audience. The room became very quiet. I waited a few more seconds, took a deep breath, and introduced myself, still facing away from the audience.

"My name is Rana el Kaliouby, from Cairo, Egypt, and I have just joined the Computer Lab. My topic is teaching computers to read human emotions, especially from our faces."

With my back still turned to the audience, I then launched into my presentation. "People express their mental states all the time, even when interacting with machines. These mental states shape the decisions that we make, govern how we communicate with others, and affect our mental performance." Although the audience could hear me, they could not see my face. Nor could I see any of theirs.

I had rehearsed these lines scores of times, yet I felt lost. I was literally talking to a wall, staring at a blank white space. Remember how difficult it is to carry on a conversation with people in the backseat when you're driving a car? You find yourself searching for their faces in the rearview mirror, and you're constantly fighting the urge to turn around and look at the person you're speaking to. It just doesn't feel natural to continue a conversation without seeing the other person. Well, that's how it felt to be speaking in front of that crowd that morning, without any feedback. It was impossible to figure out how to pace my talk, modulate my voice, or do all the things we normally do subconsciously when we're speaking in front of a group and responding to their reactions.

I must have gone on with my talk for about a minute or two, but it seemed interminable. When I just couldn't stand it anymore, I said, "So, by now you see how important the face is in communication, and can imagine how hard it would be to understand each other if we didn't have access to it."

Then I turned around and flashed a big smile. As I scanned the crowd, all eyes were on me. No one was bored, no one was dozing, and no one had gotten up to leave. Based on the "silent conversation" in the room (amusement, interest, curiosity), I knew that I had made a strong impression. At the very least, they wanted to hear more.

Okay, this was a gimmicky way to grab the attention of the crowd, but it worked. More important, it drove home the point that when we are on our computers, we are *always* face blind. There is no opportunity to "turn around" and see the reaction of the people with whom we are speaking or to gauge their response so that we can behave appropriately. The interaction is always weird and unnatural.

As I continued my talk, it felt effortless. Afterward, I took questions and comments from the audience. Most were about the technology—this was, after all, a group of computer scientists—but one comment seemed to come out of left field.

A fellow PhD student sitting in the back, said, "Rana, you should really look into autism. My brother has autism, and he struggles with understanding nonverbal communication, especially facial expressions. I think what you're doing could be very helpful to people like him."

I had never even heard of autism before. My colleague described how his brother not only had difficulty understanding facial cues but, to add to his challenge, had an aversion to looking directly into people's faces. I was intrigued: Were there human beings born emotionally blind, struggling with the same "human" skills I was trying to teach computers?

I spent the night digging up everything I could find on autism, and was shocked by how much material there was. The word for *autism* in Arabic is *al tawahod*, which literally means "alone." Nobody used it in the circles in which I traveled. In a culture where everyone is obsessed with "What will the neighbors think?" you are not as open about family members who deviate from what is considered "normal." At that time in Cairo, it was extremely unlikely that someone would announce in front of a group of colleagues that he had a brother who was labeled "different." Furthermore, although things have improved somewhat since then, few social services are available there for children with special needs. Even today, people on the autism spectrum are not mainstreamed in classrooms the way they are in the States, and for the most part, neurotypical children don't really interact with them.

The more I learned about autism, the more I could see why Rob had pointed me in this direction.

One of the foremost experts in the field is Simon Baron-Cohen,

PhD, the head of the Autism Research Centre at Cambridge. I stumbled upon the "Reading the Mind in the Eyes" test, a diagnostic tool Baron-Cohen designed to assess the degree of autism in children and adults on the autism spectrum. The test is made up of a series of photographs of only the eyes and eyebrows of men and women, each shot reflecting a different emotional state. From these photos, the test taker must decode the mental state depicted by the expression of the model in each picture.

For example, studying just the eyes and brows, the test taker must choose from among four options: whether the person is irritated, sarcastic, worried, or friendly. This may sound like a no-brainer—how can you not tell the difference between sarcastic and friendly? The fact is, each facial expression taps into pretty much the same muscle groups. The difference between even seemingly opposing moods can be very subtle: a slight lift of a brow or an almost indiscernible squint of the eye. What makes this test particularly challenging is that you don't have the benefit of seeing the entire face (the mouth, the full nose) or body gestures, or hearing the person's vocal tones. It's tough. Try it out for yourself; the test is online (autismresearchcentre.com/arc_tests). Some people do better on it than others. I did well, perhaps because of my earlier training as a face reader.

Was there a way I could adapt the test as a training tool? Could I teach an algorithm to pass it? (Not so easy, I found out. I tried and tried, and failed.) I had never seen anything like this before, and I knew I had to learn more about Baron-Cohen's work.

Today, if faced with a similar situation, I would first check LinkedIn to find a mutual acquaintance to make an e-introduction, and then fret about the best way to approach Baron-Cohen. This might have taken weeks. As a lowly PhD student, not even in the psychology department, requesting a meeting with a renowned scientist was, in retrospect, a bit gutsy. But I didn't know any better. I dashed off an email to Baron-Cohen describing what I was doing, and he responded pretty quickly. Sure, he'd like to hear about my work, and we set up a date to meet.

The Cambridge Computer Lab, with bare walls and wide-open spaces, was state-of-the-art. In contrast, the eighty-year-old building that housed the psychology department had an almost quaint feel, from the wrought-iron gates with their decorative latticework to the brick façade and cramped lobby. The interior was packed with people, brimming with humanity.

Baron-Cohen's office was smaller than I had expected, and looked even more cramped because he is rather tall. Soft-spoken, intelligent, and thoughtful, Baron-Cohen was open and friendly with me. He speaks in a measured, modulated tone, in a calm and reassuring manner. (He is nothing like the public persona of his celebrated first cousin, Sacha Baron Cohen, the comedian and actor known for his outrageous stunts.)

I had summarized my thesis in my initial email to Baron-Cohen, and I now filled in the details, describing my vision for the "Mind Reading Machine": I was relying on the Cohn-Kanade data set, featuring six basic emotions, to train our computer to see the human "face." I knew that this was woefully inadequate in terms of teaching a machine to understand the full range and complex emotional palette of human beings.

Then I admitted something to him that I could barely bring myself to say out loud: I was stuck.

If I had not heard about autism, or visited Baron-Cohen that day, I would never have known that his group was building its own database of actors depicting specific emotions to develop tools for children on the autism spectrum. In fact, they were working on the flip side of what I was trying to do with computers. They were creating software to teach autistic children to "mind read," that is, decode the emotion and mental state of others by using nonverbal cues such as facial expressions and tone of voice.

Although Baron-Cohen called it a taxonomy, or encyclopedia, of emotion, he recognized that the face portrayed more than just purely emotional signals. It also conveyed behavioral and mental states—like fatigue, boredom, and confusion—that might not necessarily be considered emotions, but nonetheless are communicated nonverbally and required an appropriate response.

Moreover, Baron-Cohen's database included a whopping *412* emotional and mental states.

As it turned out, earlier in his career, Baron-Cohen had found himself in a situation similar to the one I was facing. Back then, the gold standard of teaching emotion cues to autistic children consisted of flash cards depicting static pictures of people displaying highly exaggerated versions of Ekman's six basic emotions: ear-to-ear smiles and overdone, bulging lower-lip frowns, and so on. These cards were more like caricatures of emotions than real faces, bearing little resemblance to the actual people and expressions these children would encounter in the real world. And as a teaching tool, they weren't working well.

No wonder: After all, the point of the training was to prepare kids to interact with other living, breathing human beings. How many people walked around looking like, well, the Joker in *Batman*?

To Baron-Cohen, the solution was obvious: Why not build a video database of *real people* displaying a wide range of *authentic* emotions, subtler ones, with all the complexity of real human faces and facial movements, just as one would experience in the real world?

Nothing like this had ever been attempted before. The first and perhaps most critical step was determining which emotions and mental states should be included in the database. The question was, where do you draw the line? There are scores of ways to express even a single emotion or mental state. For example, *fatigued* has a long list of synonyms, from *slightly tired* to *exhausted* to *worn* to *sleep-deprived*, and yet each one conveys a different meaning. Likewise, just think of the variety of terms we use to describe "love" and their different shades of meaning: *familial love, romantic love, deep affection, attraction, lust, infatuation, fondness, besottedness, tenderness, en-*

dearment, emotional attachment. Now consider the fact that the meaning of *love* changes depending on who is expressing it, from the love of a parent toward a child to the love felt by a partner in a sexual relationship.

How do you distill the vast and complex language of emotion down to the key words that really matter most? Working with a lexicographer (a person who compiles dictionaries), Baron-Cohen's group identified every single word in the English language used to describe emotion; as you can imagine, there are thousands of them. After factoring out synonyms and "similars," those words that are very close in meaning, Baron-Cohen's group settled on 412.

The 412 emotions and mental states were then divided into 24 "families," or distinct categories of emotions, so that words describing anger would be in one file, words describing romance would be in another, words describing fear would be in yet another, and so on. Around two dozen actors (including Daniel Radcliffe, who played Harry Potter) were enlisted to dramatize these emotional and cognitive states. They came to the lab to perform their parts in a special recording booth built for the project. Acting as director, Baron-Cohen gave each actor a "script" as well as instructions on how to depict a particular emotion, such as jealousy, hate, anger, or sarcasm. Or he would tell the person to "look confused" or "interested."

The actors varied by age, race, and gender, so that the videos reflected the diversity of people whom children would encounter in everyday life—young people, old people, people of all complexions and hues and from every country imaginable—a far cry from simplistic flash cards.

Each video was then reviewed by a panel of ten judges (psychology students), who rated it based on whether it captured the appropriate emotional state. If all the judges gave it a thumbs-up, the video made it into the database. If not, it was discarded. Ultimately, six different actors performed six different takes on each of the 412 emotional states. Three years after our meeting, in 2004, Baron-Cohen's group published the videos as the *Mind Reading* DVD, an interactive computer-based guide to reading emotions based on our faces and voices. It was an amazing accomplishment in both its originality and execution.

As I listened to Baron-Cohen describe the methodology behind his database, I was struck by the sheer enormity of the project he had undertaken. Although they were still in the process of building the database, it already contained significant content. I felt a bit overwhelmed. If I hoped to train a computer to understand emotion, I would need to have as diverse a database of real samples from a wide swath of the population. But as a PhD student, I didn't have the funding or the time to put one together.

For a few seconds, I felt deflated: Would I be able to pull this off after all?

And then Baron-Cohen said the magic words that any research scientist hopes to hear: "Would like to see our database? You might find it helpful with what you're doing."

He then summoned another graduate student from his group into his office, Ofer Golan, the first Israeli I had ever met.

Baron-Cohen told me years later that the magnitude of the moment—an Israeli working with an Egyptian—was not lost on him. "Here were two scientists on a similar quest, from countries that have historically been in conflict. But what I really enjoyed was watching these two scientists connect as human beings and talk about concepts like empathy. It was quite satisfying to bring people together from such very different communities and find common ground."

Baron-Cohen saw how each project could work synergistically with the other. "For example, if we could identify what features the algorithm is learning from to identify one emotion from another, perhaps that could be translated to teach an autistic person to do the same," he said.

From then on, I had access to his database, the raw video files that had been meticulously collected and professionally vetted, and that captured the full breadth of emotion. It was beyond my wildest dreams, and amazingly generous, the kind of sharing of knowledge that propels research and society forward.

I downloaded the videos and devoted the next few months to viewing every image, every actor, every emotion—and there were thousands of them. The exercise proved to be absolutely essential. Before I could teach a computer the nuances of facial expressions and emotion, I needed to develop a deeper intuition myself about how human beings decode these subtle signs. I was the one, after all, designing the training algorithms. If I were deficient in my own knowledge, I would risk creating flawed tools.

Given how vast the database was, I had to narrow down which emotions I would choose for training the algorithm. I couldn't manage all 412 in three years. I was still at square one, with zero emotion recognition. I therefore decided to focus on six different categories of mental states: agreeing, concentrating, disagreeing, interested, thinking, and unsure. Once the algorithm had mastered these, I could move on to others.

I chose these categories because each one of these cognitive states is essential to both human communication and better human-to-machine interface. For example, when you're carrying on a conversation with another person, whether in the real or the virtual world, you need to know if what you're saying is being understood by the other person; whether you're boring them to tears; or whether the other person agrees or disagrees with you so that you can respond properly. That's fundamental to EQ.

Similarly, if your computer is spitting out information that you don't understand (as happened to me when I was apartment hunting) or that is frustrating you, it should be able to alter its response, just the way a person would. Think about how different an online classroom would be if the software could adjust its learning style to fit the needs of the students.

Looking back, I see now that I set the bar pretty high for myself. These mental states are very complicated in terms of how we express them; they're not always obvious. For example, you can simply nod your head yes to show agreement, or smile, but a smile doesn't necessarily mean that you *agree* with the other person. For example, when you "beg to *disagree*," you might have a smile on your face. So, the algorithm, like a human being, had to recognize and understand these subtleties.

It took me a few months to get through that entire database. Every few days I would pick a category and watch all the videos in that category, making notes about what distinctive facial expressions I was seeing for each mental state. Often times, I would fall asleep on my laptop only to wake up a few hours later and head to bed. I was heads-down focused.

In the late fall, I was deeply immersed in my work when Wael emailed me that he was planning to visit Cambridge for a couple of days on his way to a meeting in the United States. I was thrilled; it would be our first reunion since September. His visit coincided with Ramadan, the holiest month of the Muslim calendar. During Ramadan, Muslims are required to fast from sunrise to sunset, during which time we engage in self-reflection, prayer, and reconnecting to faith. Although we still go to work and carry on with our lives, we are prohibited from drinking, smoking, engaging in frivolous activities, and having sexual relations. When the sun sets, life goes back to normal until the following dawn.

I left the lab early on the day Wael was to arrive and went home to wait for him. Nice Egyptian girl that I am, for Wael's visit, I had planned a special dinner for Ramadan to break our fast and had done all I could to set a beautiful table and perk up my dreary living room.

Wael arrived midafternoon; the minute we saw each other, we switched back into honeymoon mode. We had intended to wait until sunset before leaping into each other's arms, but within minutes, we were overcome with passion. We were all over each other, and that was that. Over dinner, we talked for hours. I filled Wael in about what was happening at the lab and the slow progress of my work, as well as the skepticism I was encountering among some of my peers. He encouraged me to keep going, and he filled me in on what was happening at his company, which was growing by leaps and bounds. Despite the months apart, we were closer than ever. I felt loved and cherished, and I think he did as well. Through the years, we had many subsequent reunions, some of which I would prefer to forget, but this first one remains a precious memory of everything that was good about our marriage.

When Wael left two days later, I dove back into my research.

9

The Challenge

Baron-Cohen offered a novel approach to autism that altered how I viewed the world and my work. For the most part, up until his research, autism was perceived in a very binary way—either you were autistic, which meant that you had a "neurological condition," or you weren't. But Baron-Cohen theorized that autism wasn't an all or nothing proposition. Rather, human beings fell somewhere along a spectrum.

At one end of the spectrum are the empathizers, those with the ability "to identify another person's emotions and thoughts, and to respond to these with an appropriate emotion. The empathizer intuitively figures out how people are feeling, and how to treat people with care and sensitivity."

At the other end of the spectrum are the systemizers, those who are driven to "analyse and explore a system, to extract underlying rules that govern the behaviour of a system; and the drive to construct systems."

Empathizers excel in human-to-human interactions, but they might not be as adept at technical skills such as engineering and math. In contrast, systemizers tend to be great at technology, numbers, and logic, but may fall short on their people skills. Based on Baron-Cohen's theory, autism would be at the extreme end of the systemizer side of the spectrum, lacking in social skills.

Most of us fall somewhere in between these two extremes. Even if someone isn't diagnosed with autism, they may still have autistic tendencies—that is, they may have a lower EQ than somebody who skews closer to the empathizer end of the spectrum, while still being able to function fully in society. Likewise, someone who is good at math may still have good social skills, which might move them more toward the middle of the spectrum.

Based on his research, Baron-Cohen found that men tend to fall more toward the systemizer end of the spectrum, while women fall more toward the empathizer end. Given that, it is not surprising that more boys are diagnosed with autism than girls; depending on which statistics you believe, the ratio is either two to one, three to one, or as high as four to one.

What I found most compelling about Baron-Cohen's theory is that where an individual lands on the spectrum is not static. For example, an empathizer who is experiencing a great deal of stress may not be as attentive to the people with whom she is interacting. Under those circumstances, she could be shifting toward the systemizer end. If a systemizer falls head over heels in love, she may work harder at being responsive to the needs of her partner and find herself shifting more toward the empathizer end of the spectrum. I was really struck by this concept of a fluid spectrum, and Baron-Cohen said that, with a bit of support, one could train systemizers to move up the empathy scale, and vice versa.

Working with Baron-Cohen opened my eyes to the true potential of Emotion AI. We may all be created equal, but we are not equally good or consistently good at everything. Some people are born with a high EQ. Others, including those diagnosed with autism, struggle with EQ. And the vast majority of us fall somewhere in between.

Cultural differences, or biases and ethnic stereotypes, can also cloud our perception and judgment. Some people, due to medical conditions like stroke, brain injury, impaired hearing, and vision loss, lose the ability to process emotion. There are times when we all may experience the disorienting feeling that we're speaking to a wall, or that we are out of our depth in dealing with an emotional situation or person. At some point in our lives, many of us could use an "emotion prosthetic" to help us maneuver through a difficult time and get a better handle on our emotions and the emotions of others.

I believe that technology can augment human potential. Just as people use canes or wear glasses or use hearing aids to help them walk, see, or hear, an emotion prosthetic can help boost our empathy skills. Such a tool doesn't take away from our other strengths, but adds to our innate abilities.

I spent so much time working with Baron-Cohen's group and attending psychology department lectures that Peter Robinson thought I had gone AWOL. One day he pulled me aside, exasperated. "Rana, you're distracting yourself by going to all those things. Just focus on what you need to build." Pushing back against my adviser was not easy for me, given how I had been raised, but I stuck to my guns. I knew that what I was working out wasn't simply an "engineering" problem. I had to study how humans process emotion before I could build a machine that could.

By nature, Peter is a skeptic, someone who has seen and heard it all, and who has grown impervious to the over-the-top hype of inventors and entrepreneurs promoting the next big thing. It's not that he isn't open to new concepts or off-the-wall ideas (like mine). What I've come to learn about Peter is that he doesn't believe something is real until he can see it for himself.

When Peter challenged me, I thought, *He doesn't think I can do it.* I was used to being the golden girl, showered with praise and honors.

Now I was in a place where I had to prove myself. That said, I wouldn't be as successful as I am today without his constant prodding. What made him such a good mentor was his ability to strike just the right level of involvement; he was there when I needed him, but he wasn't overbearing. He encouraged me to take ownership of my work. By my second year at Cambridge, Peter and I had grown closer, and his unrelenting support altered my destiny. He made it clear that if this software was going to be useful in the real world, it had to work in the real world. That meant that the algorithm had to perform outside the lab, under real-world conditions, where the environment (like lighting) was often out of my control. That made the task more difficult, but it made me a better scientist.

TRAINING AN ALGORITHM

Training an algorithm is a lot like training a dog. Both tasks require endless patience, repetition, and positive reinforcement. Suppose, for example, that you're trying to train a dog to play fetch, a simple enough game. Few dogs would intuitively know that when you throw a ball, you expect them to retrieve it and return it to you. You first have to train the dog to perform this function, and the most effective way to accomplish this is to break the big goal down into smaller tasks.

First, you show the dog the ball or object to be retrieved. Next, you throw the ball and run after it with the dog. And then, when you catch up with it, you put the ball in the dog's mouth. You pat him on the head, tell him, "Good dog," and give him a treat. And then you do this again. Eventually, your dog puts it all together, and when you throw the ball, he dutifully retrieves it.

Similarly, if you're training a math algorithm, you don't overload it with a lot of instructions at one time. As with training a dog, you start small, and keep layering on different levels of difficulty. An algorithm is a set of instructions essentially telling the computer to "do this." In my case, the goal was to train my algorithm not only to recognize faces and discern features but to identify facial expressions and infer meaning. That task required a number of interim steps. The computer had to first "see" the face, identify the features, analyze the expression, and then make a judgment call based on probability. This is one approach to AI building called "machine learning."

Each step was difficult and time-consuming. For example, I first had to teach the algorithm to "find a smile." To do that, I had to feed it lots and lots of smiles from the database, and then quiz the program to see if it had learned the lesson. And as the algorithm mastered each step, it, too, would get a reward. There are actual "reward functions" (equations) built into the algorithm that give it brownie points for getting the right answer and deduct points if it gets it wrong. (Think of it as "good algorithm," "bad algorithm.") The algorithm strives to accrue as many brownie points as possible.

Now, here is where the jargon can get a bit misleading. When I talk about "feeding an image to an algorithm," I'm not showing it an image that a human would recognize as an image, like a face. Rather, I am communicating with my algorithm in a language it understands, breaking the image down into pixels and assigning numerical values to each pixel. The number of pixels depends on the image's resolution, but let's assume there are 96 pixels packed into one inch. If that pixel is black, it would be given a value of 0. If that pixel is white, it would be given a value of 255. But if the area being rated is in between, grayish, it would get a value in between—say, 125, depending on the shade of gray. So, you end up with a list (or matrix) of numerical values, and that is what the algorithm "ingests" as input. It breaks down a smile, say, into its basic components, pixel by pixel, number by number: the contour of the mouth, the lift of the lips, the crinkles around the eyes, and so on.

As the algorithm is exposed to an increasingly wider variety of smiles on a diverse number of faces that vary by gender, age, and race, it learns through experience. Over time, the more smiles the algorithm is fed, the more smile-savvy it becomes.

This is where "machine learning" kicks in. After a while, through "life experience," a "worldly" algorithm can look at an unfamiliar face and make an assessment: "That's a smile." It can also determine if the smile is wholehearted or weak, and give the weak smile a lower rating than a big, authentic grin.

This produces a massive amount of code, hundreds of thousands of letters and numbers on a page, and can therefore be very unwieldy. But code can be organized into a more manageable form. Blocks of code can be linked to specific sections. Think of it as a book divided up into chapters, pages, and even paragraphs.

It's an extremely labor-intensive, difficult task to produce the kind of algorithm I needed, but there are no shortcuts. There is an expression in the AI world, "Data is king." That's because you need volumes of data, and the right kind of data, to create smart algorithms. For example, if I'm teaching a child about apples and I show her only red apples, she might not recognize a green apple as an apple. I need to show her examples of different types of apples, in all different colors (yellow, red, green and red) as well as different shapes. It's exactly the same with machine learning: your algorithm is only as good as the examples you feed it.

And in the case of facial decoding, if I show the algorithm only faces of middle-aged white men, when it reads the face of a brown Egyptian girl, it may not be able to identify it—it may not even recognize it as a face at all. You could end up with a stupid, naïve, or biased algorithm. So, an algorithm is only as open-minded and smart as the human who built it.

I was working well into the night trying to train my algorithm while at the same time training myself in the nuance of emotion. At times, the work was tedious, relentless, and thankless. I felt physically and emotionally exhausted by it all. In March 2002, I hit a low point. I wasn't making much progress, and I wouldn't be seeing Wael or my family until the summer, which seemed an eternity away. I longed to be back with my family; I was suffering from acute homesickness.

On his way back from France on a business trip, Uncle Ahmed, Wael's father, stopped by Cambridge to visit me. It was a bank holiday, school was closed, and it was a glorious early spring day. Uncle Ahmed and I took a stroll around the River Cam, and when he asked me, "Rana, how are you doing?" I dissolved into tears. I confessed that I had been feeling awful; I was lonely and my work was just creeping along.

Uncle Ahmed looked very concerned and said, "Rana, you don't have to stay here if it's making you miserable. You can fly home with me tomorrow; no one will think less of you. Wael and your family will be happy to have you back."

His offer was so tempting. In my head, I was packing up the few possessions I had brought with me and imagining how good it would feel to be home in time to have dinner with Wael tomorrow night. I could put all of this behind me; Wael and I could pick up where we left off, and maybe even take a long weekend vacation at the beach. I nodded yes, and for a few minutes, we planned my departure.

Then reality hit. Could I really just walk away?

"I can't just leave tomorrow," I said, "Peter's on holiday, and I really should wait until the lab opens on Monday to say goodbye to him. I owe him that much."

So the plan was that I would wait a few days so that I could say goodbye to Peter. And then, of course, the next day I thought it over and said to Uncle Ahmed, "June isn't that far off. Let me at least fin ish the term." And then once I finished the term and returned to Cairo for a few weeks, I thought to myself, "Well, the hardest year is over. I really have to come back and finish what I started."

Uncle Ahmed was disappointed that I didn't come home; he was a man of the world and perhaps realized that I was headed on a path that was going to be very difficult down the road. I believe he had my best interests at heart. But the same ambition and determination that enabled me to get on that plane in September and shake up my life made it impossible for me to quit.

Uncle Ahmed came to visit several more times when I was at Cambridge but never again asked me to drop out.

10

Learning Human

n Egypt, once you are married, your parents and in-laws start bombarding you with the "Okay, when will you make me a grandparent?" question. So, it was not entirely unexpected when, the day after our wedding, my mother started badgering Wael and me about having children. We both loved children, and we knew that we wanted them, but we agreed that the sensible course of action would be to hold off on starting a family until I had completed my PhD.

To be honest, I was hardly an expert in birth control. The kinds of articles on sex and birth control in every popular woman's magazine in the States are not considered appropriate in Muslim society. On top of that, I had an aversion to taking pills of any kind. So, I refused to take the Pill, the simplest and most effective form of birth control. I didn't know much about the other methods of contraception, so Wael and I settled on the age-old method of birth control, the rhythm method.

Things went as planned for the first year of my PhD, but during summer break before starting my second year at Cambridge, I guess my math got a bit fuzzy. Then, in early October, I was sitting on the bus headed to the computer lab when I was so overcome by nausea that I had to drag myself up to the front and beg the driver to let me out. The following day, I woke up to waves of nausea, and I decided to go to my GP. By this point, I had begun to suspect the truth. The blood test performed at the doctor's office confirmed that I was pregnant. I was in a state of denial but facing a surge of conflicting emotions: I love kids, and I was excited at the prospect of holding a baby in my arms. I was also terrified that Peter would say, "Sorry, Rana, you need to take a leave from school," and that Wael would insist that I come home. I was convinced that my PhD plans had hit a wall, perhaps permanently. And that had serious long-term consequences for my career path. Without a PhD from a prestigious school, I had little chance of securing a tenure track position at AUC. I had no Plan B.

In a daze, I returned to my apartment and told Wael the news. He, too, was happy at the prospect of being a father, but he understood my dilemma. He didn't issue any edicts, but he urged me to tell Peter right away, so we could plan for the future one way or another.

So, without skipping a beat, I went back to the lab and knocked on Peter's door. When I walked in, I burst into tears. When he discovered the reason for my tears, he smiled broadly and said, "Rana, that's wonderful news." And he quickly followed up by promising that he would support me in whatever I decided to do: If I wanted to continue with the program, he would be there for me. And if I wanted to take a leave, he would understand.

I went back to my apartment and spoke to Wael again before I made up my mind. "Rana, if you take a leave now, you will never complete this PhD," he told me. With Wael's support, I remained at Cambridge, but not without some misgivings. I was consumed with self-doubt. Will I be a good mom or too distracted by my work? Will I be able to finish my PhD with a baby to care for? How can I survive the year without my full support system, my husband, family, and friends? There was a big question about whether I could pull this off, which made me all the more driven to make it work.

I was thrown off balance for a week or two after learning that I was pregnant, but I quickly regained my footing. I took charge—I was going to do this my way. I decided I wanted natural childbirth with as little medical intervention as possible. I opted to work with Sally Lomas, a highly trained midwife who was affiliated with an excellent hospital in Cambridge where I would deliver the baby. To me it was the best of both worlds; a doctor would be on hand if something went awry, but the midwife and I along with selected family members would manage the delivery. Wael was fine with this. I took a yoga class for expectant mothers and signed up for Lamaze childbirth preparation classes, where I learned breathing techniques to relieve the pain. Without the presence of my husband or family, I was grateful to have Sally and the other expectant mothers I met in these classes for support.

In Cambridge, my decision to use a midwife instead of a doctor was well within the norm—even trendy. Many women like me were determined to find an alternative approach to what we saw as the medicalization of childbirth. It was common to have birthing balls, scented candles, and even husbands, family, and friends in the delivery room. This was not the case in the Middle East. When I called my parents and told them my plan, my father was absolutely horrified. In fact, he couldn't believe that in a modern country like England this was even allowed! In Egypt, only the poorest use midwives, who have no formal training in childbirth. It was dangerous, and the mortality rate among these women was quite high. The fact that a woman in my position who had access to real doctors would choose a midwife was inconceivable to him. Eventually, he calmed down when I described Sally's education, and the fact that I was delivering in the Rosie, the best hospital in Cambridge. But neither my parents nor my in-laws could wrap their brains around the fact that Wael would be in the delivery room, assisting with the birth. There is no prohibition in Islam against this; the opposition is cultural, perhaps born of a belief somehow that men shouldn't see women in this state. Or, more likely, men just can't take it!

Despite the yoga and the breathing, I did not have an easy pregnancy. The nausea never really went away. I would be at the lab and excuse myself from whatever meeting I was in to rush to the restroom, where I'd throw up, wash my face, brush my teeth, and then walk back into the meeting. I sometimes surprised myself at how determined I was, but underneath the bravado, I was intensely lonely. Sometimes, late at night, I'd break down and cry. It all felt so hard, and I wasn't sure how I would keep going.

LAB MATES

At Cambridge, I was able to work with some of the top computer scientists in the world, especially in cutting-edge artificial intelligence like computer vision and machine learning. It made me a better scientist. In retrospect, though, I see that some of the most valuable lessons I learned at Cambridge were not necessarily in the realm of computer science. They were about human beings, and I think that they made me into a more informed, better person.

There were few educational or work opportunities in the Middle East for people with physical or mental disabilities. Expectations for their success in life were low. I was therefore surprised, actually completely gobsmacked, to find myself sharing an office with a partially sighted young British man named Silas Brown. Silas had a condition called cortical visual impairment (CVI). His eyes worked fine, but there was a glitch in the optical processing system in his brain that resulted in very poor sight. Still, he moved quickly around the lab and campus with the aid of a white cane.

Clearly, Silas had done well in life despite his limited vision. I was impressed with his ability to brush aside his visual deficit and forge ahead. He never complained, and seemed to take his condition in stride. But he did confess to me that he often felt uncomfortable carrying on conversations with others—not because he wasn't well spoken or was socially awkward, but because his limited vision prevented him from seeing the other person's facial expressions. He was often in the dark as to whether the person he was chatting with was interested or engaged. It made it difficult to respond in an appropriate way. I empathized with this problem. My research on autism had opened my eyes to the obstacles facing people who were, for whatever reason, blind to nonverbal cues.

Silas was a Jehovah's Witness, the first I had ever met. Jehovah's Witness is a branch of Christianity that seems even stricter than Islam. Silas was forbidden from celebrating most religious holidays, nor did he celebrate his own birthday. Similar to Muslims, Jehovah's Witnesses must adhere to a strict code of conduct regarding interactions between men and women. Silas was cut from the same "modest" cloth as I was; he was definitely not a "hugger." I was amazed by the level of Silas's devotion to his faith; it made me realize that there are many different ways we can worship and express our faith.

In the fall of 2002, around the same time I learned that I was pregnant, Tal Sobol-Shikler, an Israeli PhD student, joined the lab. There has been some bad blood between our two countries, and both Israelis and Egyptians still have vivid memories of the 1973 war. Tal learned that there was an Egyptian student in the PhD program and, sensitive to this, and unbeknownst to me, wrote to Peter and said that she didn't want to apply if he thought it would cause conflict. I don't think I struck Tal as a woman looking for a fight, and we quickly became good friends.

Tal was married with two small children. Her husband was also studying at Cambridge; she wasn't on her own. I observed how she juggled her career, marriage, and parenting responsibilities, and I knew that, by this time next year, that would be me. Like me, she was ambitious, and despite having a family, she wanted to excel in her field. We had something else in common: a desire to humanize technology. Her thesis project, an analysis of affective expression in speech, was similar to mine, except that she was teaching computers how to decode verbal cues rather than facial ones.

Fall turned into the cold, dark days of winter, and my algorithm still could not identify even one single facial expression. I felt intense pressure to prove myself. After all, I had only a year and a half left to finish this complicated thesis project. If things didn't start falling into place soon, I might not make it.

I was starting my seventh month of pregnancy and tried to pack as much work into my days (and nights) as possible. I would wake up, wobble out of bed, take the bus to work—I had long given up biking—and be in the lab by nine A.M. Then I'd work until about four P.M. or so, take the bus back home, shower, cobble together something to put in the microwave for dinner, and then plop onto my beige couch with my laptop propped on my growing tummy and code away until late at night. Often, I'd have the TV on in the background, just to keep me company.

What in the world was taking me so long? It wasn't as if I were trying to teach the algorithm the entire repertoire of facial expressions. I was working on teaching it one basic expression: the head nod. I thought it was a good way to begin because a nod is a more obvious movement than, say, an eyebrow raise or a lip curl—it's hard to miss the up-and-down motion of a nod.

Still, from a machine learning perspective, teaching an algorithm to identify a nod was complicated because it involved moving parts. I couldn't use only a single frame or a static picture; my algorithm would have to learn the entire head movement, from start to finish, as it unfolded over time. And that held true for most nonverbal cues, even the simplest of expressions, like a smile. A smile unfolds over time, and that *temporal signature*, the pace at which the expression unfolds, is very telling.

Different nods have different meanings. The basic, simple up-and-down movement of the head signals consent, a "yes," or *aywa* in

Arabic. But the pace at which you nod reflects another whole layer of meaning. For instance, nodding slowly means something very different from nodding quickly. The former could be interpreted as a hesitant agreement, while the latter implies wholehearted agreement. Nodding up and down only twice conveys a very different meaning from a nod you do five or six times. My "head nod detector" needed to understand these intricacies so that it could recognize and respond to all these different forms of nods.

I spent my evenings trying to code for these temporal signatures—slow nods, fast nods, two nods, six nods—feeding the algorithm thousands of examples as part of the training process. When the algorithm analyzes the image, it spits back a number (a probability score) between 0 and 100, which measures its assessment of whether the image is a nod or not. If it spits out a number close to 0, then it doesn't think the image is a nod. Conversely, the closer the score is to 100, the more likely the algorithm thinks the image is a nod.

For months, I was getting way too many probability scores in the 50 range. A flip of a coin would have done as well! Very depressing.

Late one night, after coding the newest version of the algorithm, I fed it examples of lots of nods it had never seen before. Then I waited to see what numbers it would spit back. I was tired and ready to give up for the night when I saw the first result:

Test head nod 1: probability score 91 percent.

That was excellent.

Test head nod 2: probability score 95 percent.

My pulse started to race. This was a far better score than I'd ever gotten before.

As I continued, I began to get excited, but I didn't quite believe it yet; there were more tests to do. I decided I had to mix things up a bit, to ensure that the algorithm wasn't just naming everything a nod. So, I snuck in a few examples of a head shake, a head tilt, and random head movements. Every time I tried to trick the program, it spat back a low-probability score.

In other words, I had done it. After a year and a half of teaching it "human," the algorithm, after hundreds of hours of training, was finally able to distinguish between nods and non-nods. If I could have done an Egyptian *zaghroota*, the ululation of joy my people make at weddings, I would have.

I had finally cracked the algorithmic code for reading facial expressions. The head nod was just the beginning. Now that the algorithm understood what it needed to look for, and how expressions unfolded over time, frame by frame, pixel by pixel, I could layer in different data. I could use this same algorithm to train for smiles, frowns, raised eyebrows, squints, and crinkles—the full range of facial expressions. It meant that I would be able to finish my thesis. I would be able to build the Mind Reader.

JANA

I continued my work, and practically coded my way into the delivery room, working right up until a few days before my daughter, Jana, was born. I felt my first labor pain in the late afternoon on May 23, 2003, at my final breathing class, with Wael present. We went home until eleven P.M.; later, when the contractions were eight minutes apart, we headed to the Rosie Hospital, where we met Sally. My mother, my sister Rasha, and my mother-in-law had all come to Cambridge for the birth; my mother joined Wael in the delivery room.

The first thing Sally did was slide the hospital bed to the wall so I had room to move around. The only medical intervention was the monitor I wore around my waist to track the baby's heartbeat. Throughout the night, I bounced on the birthing ball, paced around the room, did some yoga poses (like the rocking cat, on all fours), and breathed through the contractions with Wael as my coach. My mom sat in a corner chair and read and re-read verses from the Quran. It was raining outside.

The contractions came faster and stronger, each one more painful.

I began to push. At 8:55 A.M. on May 24, Jana was born. Sally placed her on my chest, still attached to me by the umbilical cord. She looked at me with her steel gray blue eyes, and I was overcome with relief and joy to be holding my little girl in my arms. I started to cry. I felt both gratitude and awe; I vowed to be the best mother that I could be.

As is the custom in Islam, the first thing a newborn should hear is the Iqama, or the Muslim call to prayer. My mom, aware of this, leaned over immediately following Jana's birth and whispered the prayer softly into Jana's ears.

> Allaho Akbar [Allah is the greatest];
> I bear witness that there is no God but God
> I bear witness that Muhammad is the Prophet of God
> Come to prayer
> Come to success
> God is great
> There is no God but God

Years later I learned that it is customary for the newborn's dad or grandfather to whisper the prayer, rather than a woman! But my mother is a devout Muslim, and it just felt right that the matriarch of the family was the one to introduce her granddaughter to her faith. Little did I know we were already breaking down gender roles.

Ten days later, Wael, his mother, and my sister returned to Cairo. My mother stayed on to help out with Jana for a few more weeks. I was back at the lab a week later—not because I felt pressured to return to work, but because I wanted to be there. The research was going well, and I wanted to maintain the momentum. I spent the time putting the finishing touches on a paper describing my work that a few days later I submitted to the International Conference on Intelligent User Interfaces, a major gathering of computer engineers. I didn't know whether the paper would be accepted or not, but I was determined to send it in before the deadline.

In August, I took Jana, now three months old, on her first flight, back to Egypt to visit family. It was a five-hour British Airways flight. I was breastfeeding her at the time but still wasn't comfortable doing so in public, which meant that I hesitated to feed her on the plane. And even though I had prepared bottles of pumped breast milk ahead of time, Jana wouldn't drink from them. She must have sensed my anxiety, because she screamed and cried the whole time. I did my best to comfort her, but it was a traumatic flight for the two of us. The minute we set foot in Egypt she calmed down; I was very happy to be with my family, and everyone fussed over her. She was completely spoiled, as was I. And after all that time taking care of myself, it was wonderful being cared for by others for a change. My friends organized a small celebration for us; the running joke was "How did you become pregnant in the first place—over the Internet?" Everyone knew that Wael and I hardly saw each other!

All too quickly it was September. I took a flight back to Cambridge with Jana. It was time to spring back into action.

11

Mommy Brain

A lot has been written about the so-called mommy brain—how, after she gives birth, a woman's cognitive abilities decline for a time as she focuses on her baby. It's true that women's brains do undergo certain changes after childbirth, especially in areas involving the so-called maternal instinct, the regions in the brain that control empathy and understanding. And I did feel a rush of new emotions, a deep and protective love for Jana that I had never felt before for any other living thing. But having a baby did not throw me off my game. In fact, my mommy brain was in high gear.

Yes, I wanted to carve out as much time with Jana as I could—my time with her was precious—but that didn't mean I was ready to give up my other dream. It just meant that I had to be even more efficient with my time. In a way, having a family in Cambridge made me even more motivated; I wasn't lonely anymore. My mother came often to help out when she could, but I still had to put Jana in daycare. I found a highly recommended daycare center a block from my apartment building. The staff was well trained and warm, and a number of the Cambridge faculty used it for their children. What I had not anticipated was the surge of guilt I experienced when I dropped off my five-month-old helpless infant. I fretted that because she was so young, she would be scarred for life and become an emotionally detached human being. My fears quickly subsided; Jana loved daycare and was a happy, responsive baby and lively toddler. As it turned out, her experience in early care seemed to make her more adaptable and accepting of new people and change in her life. Jana doesn't get thrown by new situations. She is confident and quick to adapt.

Even with childcare close by, juggling all of this wasn't easy. My day began at four A.M. I'd work on my thesis until Jana got up around six or so, then I'd feed her. We'd play a bit, and I'd drop her off at daycare. After that, I'd go back home to pick up my bicycle and on to work, where I switched right back to PhD mode. I stayed laser focused on work until around four P.M., when I'd pick up Jana and then switch back to mommy mode. I'd take her for a stroll around the River Cam or I'd read her a book, and when Jana went to sleep, I'd switch back again to work mode. There was no room in my life for socializing. I even skipped most of the group lunches at the lab. I excelled at compartmentalizing. I reasoned that the quicker I got done with my PhD, the faster I could return to Cairo and spend more time with Jana, and she would see more of Wael. Despite the long days, I knew I had a wonderful baby at home, and I felt fulfilled and productive at work.

To most people, academia suggests a rarified, ivory tower existence. But the reality is that when you're chasing a discovery, it can be as hotly competitive as the real world. I knew that researchers in other labs were working on similar projects, and I wanted mine to be the first and best out. So, I kept up the pace. Between breastfeeding and getting to know my baby, I coded.

Late that fall, I received some very good news: The paper that I had submitted to the International Conference on Intelligent User Interfaces right after giving birth had been accepted. I was invited to present the work as a "poster presentation" at the annual conference in Madeira, Portugal. A poster presentation is literally what it sounds like: I summarized my work on a 48-by-36-inch piece of poster board, which was displayed in the large hall, like at a science fair. For hours on end, I stood next to the poster describing my research and spoke with the hundreds of academics and industry professionals who came by. Six-month-old Jana was parked right next to me in her stroller. When she became fidgety or hungry, I'd duck out of the conference hall, feed her, and then return to my post. Yes, it was awkward; the majority of the attendees were men, and I was concerned that, as the only woman in the room with an infant, I wouldn't be taken seriously. But in a way, it worked in my favor. Among the two hundred or so presenters, I stood out. In fact, Jana and I were a hit: people quickly recognized the hijabi new mom and her adorable infant.

This was the first time I had presented my work in such a public forum, and the response was overwhelmingly positive. Other scientists could see that I was on to something; I felt respected and recognized by my peers. My work rocked, and that was what mattered.

I now had two jobs, scientist/inventor and mother of an infant, and I was determined not to fail at either. I worked extremely hard my third year at Cambridge. It was an incredibly productive time for me. Now that my algorithm could read human facial expressions, I was able to teach it a variety of new expressions in rapid succession.

When spring finally arrives in Cambridge, sometime around April, the days get longer and the city comes alive with greenery. On weekends, weather permitting, I would take Jana out to the parks, where we would sit among the daisies. Or we'd run and roll around in the grass together. Egypt and Kuwait are largely desert, so I loved the smell of fresh grass and the flowers, and wanted Jana to experience the open expanses of England before we left for good.

The year passed quickly. In early June, as I was wrapping up my thesis and preparing to return to Cairo with Jana for summer break, Peter Robinson circulated an email announcing that Rosalind Picard, the professor who wrote *Affective Computing*, would be touring the lab on August 24. Picard wanted to meet with some students and learn about their projects. Peter anticipated that she would give each student about ten minutes, and he urged anyone who was interested to sign up.

I didn't realize it at the time, but I had to make some difficult choices then, choices that would ultimately impact both my personal and professional life. Of course, I wanted desperately to meet the woman who had inspired me to do the work I was doing, whose book had ignited my imagination; but there was a catch: The Mind Reader needed additional work before I felt comfortable showing it to Picard. Before I had learned of Picard's visit, I had planned on going home for the summer and returning in the fall to finish my thesis, but now I was reconsidering that decision. It was a difficult choice. Wael and I had barely seen each other over the past three years. Summer was a time for us to get reacquainted, and for Jana to see her father and extended family. Little did I know that this trade-off between Rosalind Picard–related matters and my family would become the norm.

I called Wael and explained the situation to him. He agreed that I should stay on at Cambridge to meet Professor Picard. Looking back, I see that it had to have been hard for him. He must have been disappointed. I would have been if it had been the other way around. But if he was, he never let on.

For my demo, I used a big Logitech webcam attached to the lab computer and a large monitor. The monitor displayed the image of what the webcam was seeing, the face, with lots of line graphs scrolling underneath it, and green and red bars on the side. These graphs and bars provided a readout of the "face's" mental state, showing in real time whether the person was smiling or nodding, interested or confused.

I had not heard of any project at any other lab that came close to what I had achieved, but I was still worried. Picard ran the Affective Computing Group at the MIT Media Lab. MIT was a juggernaut in computer science, and I had a nagging fear that someone at

Picard's lab had already cracked the code and that my work wasn't innovative at all. With this fear in mind, I was driven to perfect what I had done. I wanted to give it my absolute best shot.

On the morning of the demo, I woke up early to give myself extra time to get ready. I had gone through my closet, mulling over what outfit I would wear for the meeting. I wanted to look confident, smart, elegant, and formal, but not too formal. Most of all, I wanted to be memorable. After experimenting with various outfits, I selected an orange top and a matching headscarf and navy pants.

I dropped Jana off at daycare, a ten-minute walk from my apartment, and then took the bus to the lab. There, I fired up my computer and started my demo, testing it over and over again to make sure it worked. I reviewed my script in my head. First, I'd introduce myself, tell Dr. Picard about my work, and then invite her to see a demo of the Mind Reader. I was on pins and needles waiting for her to drop by my office.

Picard arrived promptly, looking crisp and professional in a blouse, blazer, and pants. She had short blond hair and a face that exuded intelligence and curiosity.

After I had introduced myself, Roz—she insisted that I call her "Roz"—started firing questions at me. What emotions was I focused on? What methods had I used?

Dynamic Bayesian networks, I told her, because I want to incorporate temporal information as facial expressions unfold, as well as encode the complex mapping of facial expressions and their meaning.

What had I implemented my system in?

I told her that I had programmed it in C++, to ensure that I could build a real-time demo.

What data did I use?

I told her about Simon Baron-Cohen's database.

All the questions and answers meant nothing, of course, if the demo didn't work. With butterflies in my stomach, I invited Roz to try out the Mind Reader for herself. She sat down at my desk, looked straight at the Logitech webcam, and started making faces: smiling, frowning, looking interested, looking surprised. It worked every time. I could tell that she was impressed, *really* impressed. I breathed a sigh of relief.

Roz stayed for forty-five minutes, and I was totally in my zone. We were like kindred spirits, talking quickly, exchanging ideas, and completing each other's sentences. We clicked. I had finally met someone who fully appreciated what I did and who understood what I aspired to build. I talked about my aha moment, the realization that this wasn't just about human–computer interfaces, but about the way people communicated. And I shared my vision for creating an emotion prosthetic for autism.

Finally, Roz said, "This is incredible work. Would you like to come work for me as my postdoc after you finish your PhD?"

I have to go back to Cairo, to my husband, who has been waiting for me for three years!

I sat there quietly for a few seconds. I told Roz that her book was the reason I had come to Cambridge in the first place, that I'd followed her work for years. She was not just an inspiration, but a role model. Working with her would be a dream come true.

But I also told her, in a playful tone—and for some reason, I still remember the exact words—"I am a Muslim, and in Islam, your husband can marry up to four wives. I have been away for over three years now, and if I don't get back after my PhD, he will most definitely take a second wife. And so, as much as I'd like to join your lab, I need to go back to Egypt after my PhD."

I was half-joking. Although polygamy is still practiced throughout the Middle East, it is very rare among my educated circle. Still, I was concerned about the fate of my marriage if I stayed away again.

I didn't know it at the time, but Roz never takes no for an answer. My response only intrigued her more. After asking me about my situation, she announced, "We'll figure something out. You can always commute from Cairo."

When she left, I called Wael right away. I had met Rosalind Picard and it had gone well—really well, I told him. I didn't tell him the part about her inviting me to join her lab; I figured we'd cross that bridge later.

My work wasn't over yet. I still had to write up my thesis, and that would take several more months. I wanted to stay in Cambridge to finish up, but Wael insisted that I come home. He missed Jana and wanted me back. It was understandable on his part, but I wanted to stay and complete my work in Cambridge. In Cambridge, I had a rhythm to my life that revolved around work and Jana. In Cairo, I would be sucked back into family life and other social obligations. I was worried about becoming so distracted that I wouldn't be able to write. I pushed back and told Wael I wasn't ready to go home yet, but he put his foot down and I gave in. Fortunately, Peter understood my situation. I had put in enough time at the lab that he allowed me to work from home.

For the six months I was back in Cairo, I stuck to a rigorous schedule like the one I had established at Cambridge. I woke up at four A.M. to write before everyone else got up, and during Jana's naptimes. Right about the time I was wrapping up my PhD (in early spring of 2005), I realized that I needed continuous, uninterrupted time to finish the final chapters, and it was hard to do while caring for Jana. Wael's job as CEO of ITWorx was demanding, so he couldn't do much babysitting. So my mom took a leave of absence from her teaching job to help me. I packed up Jana and we headed to Abu Dhabi for a month. My mom would keep Jana company, taking her to parks and shopping malls, while I locked myself in my childhood bedroom and wrote nonstop.

One day, I dropped in at my dad's office for a visit. The office was filled predominantly with men, except for the few women who worked as assistants. My dad works for the minister of the interior, where he is responsible for implementing the information technology and AI systems that power the country—e.g., police stations, fire stations, airport security, etc.

I walked in and asked for Ayman el Kaliouby. One of the men called out, "Where is Abu Rana? Tell him his daughter is here."

Abu Rana? I couldn't believe what I had heard.

In Arab countries, men are customarily called by the first names of their eldest sons. So if I had a brother named Ahmed, my dad would be called "Abu Ahmed." If a man had no sons, then he was simply called by his own first name, i.e. Ayman.

My father's co-workers called my dad by the first name of his eldest daughter—me! The fact that they bestowed this title on my father meant that he must have talked about me often and was proud of my accomplishments, and that his co-workers acknowledged and celebrated the fact that my dad had a daughter worth honoring. I was deeply touched by the gesture.

VIVA VOCE

I returned to Cambridge early the next spring to defend my thesis. In the United States, to complete your doctoral program, you have to "defend" your thesis in a public forum, where you review the work you've completed and explain your approach and accomplishments. There is a similar protocol in the United Kingdom, called the viva voce (Latin for "live voice"), or "oral examination," and truthfully, I was dreading it like a root canal. It's tough and intensive; everyone hates it. You are assigned two examiners and a chair, who runs the program. My committee consisted of Roz Picard (who came at my invitation), Peter Robinson, and Sean Holden, a member of the machine learning faculty at Cambridge.

The viva voce is a closed event, and there are no rules for its duration. The examination normally lasts at least ninety minutes and may go on for up to three hours. At the time, I had no idea how long mine lasted, but I do remember that after the initial queasiness, the words seemed to flow out of me. After three and a half years, I knew my field cold. I was actually enjoying myself.

Roz Picard asked a lot of forward-looking questions: "What will you do with this?" "What are the applications?" "What are the gaps in the system?" I still had not agreed to come to the MIT Media Lab, but I could see that Roz was thinking ahead, pondering what we could do with this.

After completing my viva, I moved back to Egypt. My last couple of weeks in the lab were very tearful. For the past three and a half years, my friends at the lab had been mine and Jana's surrogate family, and I am loyal to the people I grow close to. The uncertainty of the future made me anxious.

In May 2005, I returned to Cambridge for graduation, accompanied by Wael, my parents, and Jana. It was my father's first visit to Cambridge, after all these years. Graduation took place on one of those rare, clear blue-sky days in Cambridge, with brilliant sunshine. It was picture perfect, with no fog, rain, or grayness.

I wore a cap and gown, of course, although my cap was perched atop a hijab. (I had a bit of a struggle to make it stay there.)

Cambridge is a forward-thinking institution, but one that respects the past. After all, it was founded in 1209, and many of its traditions date nearly as far back. I remember marching in with the first group of candidates for higher degrees, and I remember hearing a lot of Latin, and then walking up to the stage.

"Most worthy vice chancellor and the whole university, I present to you this woman whom I know to be suitable as much by character as by learning to proceed to the degree of doctor of philosophy for which I pledge my faith to you and to the whole university."

My name called, I stepped forward and kneeled.

"By the authority committed to me, I admit you to the degree of doctor of philosophy, in the name of the Father and of the Son and of the Holy Spirit."

Jana, who was about to turn two, attended all the commencement events and wore a cute white dress. I gave her a bottle of bubbles to blow, to keep her entertained and avoid any tantrums. As I went from event to event, she stayed close by my side, giddily blowing bubbles all around me.

Here's what I remember most about that day: For this one precious moment, it felt like I could have it all: an amazing career, a supportive husband and family, and a healthy, happy daughter. I knew I was lucky. Not everyone gets to experience that feeling in their lifetime.

12

Crazy Ideas

f the theory of parallel universes is true—and who am I to argue with Stephen Hawking?—somewhere in space and time is a Rana who moved back to Cairo, became a tenured member of the computer science department at AUC, lives close to the campus in a stylish suburban home that is fully "wired" but featuring the best in contemporary Arabic architecture, and is still happily married with, who knows, maybe three or four kids by now . . .

That was the "grand plan," the life I always believed I would fall back into after I earned my PhD. But the stars didn't quite align that way for Rana on planet Earth.

Nonetheless, in the fall of 2005, when I returned from Cambridge, now Dr. el Kaliouby, I was still "on track," teaching CS106 (Introduction to Computer Science) at AUC, my alma mater. At twenty-seven, I was one of the youngest members of the AUC faculty, and I strived to bring fresh, new thinking to my classroom. When I took the same course more than a decade earlier, the focus had been solely on programming. But to me, it wasn't enough to teach my students the art of coding. Even though this was before the iPhone and other smartphones, more and more people were using Myspace, AOL Instant Messaging, Google, and chat rooms, and were shopping sites like eBay and Amazon, leaving a mountain of data about their behavior with every cyber interaction. To those of us in the field, it was clear that data was the new currency, and would only increase in value as technology gained greater insight into how we lived and into our preferences, concerns, medical conditions, and the like. Now computer scientists had to consider issues that were never before relevant to our work. (Protecting the privacy of users and not breaching their trust by selling their personal data to the highest bidder was not a major problem before the formation of vast databases.) Moreover, I wanted students to think about creating products not just for an elite few (the educated, well-off, and able-bodied), but for everyone, from all spheres of society.

I therefore engaged my students in deeper discussions, tackling issues regarding the ethics and moral responsibility of tech companies, something rarely taught in computer classes. Perhaps if ethics had been a mandatory part of the core curriculum of computer scientists, these companies wouldn't have lost the public trust in the way they have today. I wanted to inspire these kids to use their knowledge to make a positive impact on technology, and perhaps even change the trajectory of their lives.

I liked teaching, I loved AUC, and there was something, well, so comfortable about being back at the school where I had spent four of some of the happiest years of my life. And after so many years apart, it felt wonderful being home with my husband, surrounded by family and close friends whom I had missed every day I was away. I had left the drab gray skies of England behind. The days were almost always bright and sunny in Cairo—no need to check the weather forecast anymore.

My life was, as the Brits would say, "posh" in comparison to how I had lived in Cambridge. No longer did I pedal my way to the office, or stand on a cold corner waiting for the bus. We had a driver, Salah, a Nubian (an ancient people indigenous to Sudan and southern Egypt), who drove me to work every day in a BMW. Traffic is terrible in Cairo, so we spent many hours together in the car talking, and Salah and I are still in touch today. I had a full-time housekeeper and a live-in Indonesian nanny, plus family members ready and eager to babysit. I didn't have to shuttle Jana to daycare, cook a meal, or lug a bag of laundry to a coin-operated Laundromat ten minutes away. (Late at night, when I'm throwing in a few loads of wash and prepping vegetables for the next night's dinner, I often think back to those days in Cairo.) Every Friday, on the Muslim holy day, Wael would pray at our mosque, and then we would all have a leisurely lunch with my in-laws.

When I returned from Cambridge, Wael and I rented a house in Al Rehab, a very desirable, exclusive new community being developed on the outskirts of Cairo to deal with overcrowding in the city, which was bursting at the seams. In 2008, AUC was slated to move from downtown Cairo to New Cairo. As a faculty member, I had the option to purchase a parcel of land to build a house near the new campus at a significantly reduced price. It was a prime neighborhood, and Wael and I grabbed a choice plot and planned our dream home. Although my affiliation with AUC gave us the opportunity to purchase the land, I relinquished all my property rights to Wael. I just assumed that he and I would be together forever. As is custom in the Middle East, I left all the financial and legal details up to him. It never dawned on me to ask any questions, or to insist that I owned a piece of the land and the house. I just signed the papers. Nor did I ever ask questions about our finances. That was a piece of my education that was sorely lacking.

Still, we enjoyed a lovely life. I was hardworking and conscientious about pursuing my career, but I was also pampered in a style that Westerners can't understand. Help is cheap in the Middle East, and a well-to-do middle-class person can have some of the trappings of luxury that a much wealthier person in the States enjoys. Perhaps if I had never been to Cambridge or met Roz, I would have been content to live that way for the rest of my life. (But of course, if I hadn't gotten my PhD from Cambridge, I would never have been hired by AUC.)

Ever since Picard invited me to join her Affective Computing Group at the MIT Media Lab, it was always lurking at the back of my mind. Part of me didn't want to contemplate the possibility of accepting the postdoc. (*If I do get a postdoctoral job, how will I balance this with family and Wael?*) But, somehow, I couldn't let it go, either. And neither could Roz, who was determined to bring me to her lab and was cooking up ways to make that happen.

Wael knew that Roz wanted to bring me to MIT, but he and I never really talked about it in any depth. Given his negative reaction when I asked if I could stay in Cambridge to finish my thesis, I thought it best not to even bring up the possibility of my leaving again. I wasn't even sure that it would happen.

Roz didn't have the funds to hire me that year, though I was willing to work for free for the opportunity. But as a foreigner, I needed MIT to sponsor my visa, and that meant a more formal arrangement, with a salary.

So, Picard decided that she and I should apply for a National Science Foundation grant in computer and information science and engineering. If we won the grant, there would be money to bring me on board. But the NSF grant was hardly a sure thing. It is tough to get, and the application itself is daunting.

We requested a grant to build the "Social-Emotional Prosthesis for Autistic Individuals," a new tool to enable people on the autism spectrum to better understand the emotional cues of others. This would bring the technology I had developed with the Mind Reader into the real world in exactly the way I had envisioned it would be used: to enhance human-to-human communication.

We were proposing to embed a camera on a Google Glass type of device and program it to identify facial expressions in real time and give feedback to the wearer via ear buds. Well, in 2006, there was no Google Glass; iPhone had yet to be launched and cameras were not yet so ubiquitous or advanced. Our "crazy idea" was not only ambitious but definitely in the realm of "out there."

But then again, so was my Mind Reader.

Corresponding from afar that fall and into winter, Roz and I worked furiously on the grant application, which was lengthy and detailed. I would write in the morning, Cairo time, then email the draft to Roz as Boston was waking up. Roz would then work on it while I was sleeping in Cairo, and email a draft by the morning (my time) the next day. We had a running joke that we worked around the clock.

Our proposal was well crafted and thought out, and Roz and I worked as well together as I had hoped. Even better, I had shown her that I could work remotely and be effective despite being halfway around the world. I hoped that if all went as planned, she would be amenable, unlike Peter, to my working from home a good deal of the time.

We submitted the grant in the late fall of 2006 and heard back in early winter. It was the most positive rejection I had ever received. The National Science Foundation loved the idea. Our proposed project had scored high on "potential impact," and they felt that the proposing team (Roz and I) had the intellectual merit needed to actually build the thing. But after all that, they deemed the project way too ambitious, an impossible undertaking, and concluded that it would never get built.

Too ambitious? Impossible? Really? Oh, I'd heard that one before, and it hadn't stopped me back then.

Still, I didn't see a way forward. I was sorely disappointed by the rejection, and I knew that the next step for the Mind Reader (my algorithm) was going to be tough. It would require the mentorship of someone with the skill set and vision of a Roz Picard—and the truth is, there is only one Roz Picard in this universe. And we really couldn't collaborate effectively unless I was part of her group at MIT. For a time, it seemed that my dream had withered on the vine.

Shortly after receiving the "You're great, but no thanks" email,

Roz sent me an email asking me to call her. I braced myself for her response to the NSF rejection. I played it over and over in my head: "I'm so sorry, Rana, we tried; it didn't work. Good luck with your life." I would be grateful for her help and try not to be too disappointed.

I finally gathered up the courage to make the call. My hands were shaking; my heart was pounding. I hated to think that this dream would soon be over. Roz picked up right away, and her voice was surprisingly upbeat. The first words out of her mouth were, "Okay, Rana, they love the idea; they just don't think it's doable. So, let's build it first and then apply for the grant again—a larger grant next time!"

I was stunned by Roz's determination; clearly, this woman doesn't give up. I was heartened that Roz was still figuring out ways to get me to the lab, and I appreciated her enthusiasm. But I was still a bit wary, and I didn't want to get my hopes up again.

"How are you going to make this all work?" I asked. I didn't doubt Roz's sincerity, but from my perspective, the odds seemed overwhelmingly against us.

"No worries," she said confidently, "I'll talk to Nicholas Negroponte."

I learned a great many life lessons from Roz, but perhaps the most important is that persistence is a key ingredient to success: Never take no for an answer, and never underestimate Roz Picard.

Negroponte, a revered name in technology, founded the iconic MIT Media Lab in 1984. He is famous for many things, but especially for infusing a more humanistic view into technology. He is often quoted as saying, "It's not computer literacy that we should be working on, but sort of human-literacy. Computers have to become human-literate."

If anyone understood, and would champion, what Roz and I were doing, it was Negroponte. At the time, he was in the process of leaving the lab to focus on his One Laptop per Child initiative, a nonprofit with the mission to distribute low-cost laptops to children worldwide to promote educational opportunities for all. Although he had one foot out the door, Negroponte still held some sway over how the lab's budget should be allocated. Before he had shut the door for good, Roz approached him and asked for money to fund someone the lab had never seen before: a young Arab-Muslim Egyptian woman and Cambridge PhD who wanted to build "an emotion prosthetic."

I waited on pins and needles to hear my (our) fate. At the same time, life went on. I had returned home just in time to help with the planning of my sister Rasha's wedding. Given the more conservative tone of the times, this wedding wasn't going to be the kind of big, loud, brash celebration that Wael and I had had, with a deejay, belly dancer, and singer. That was no longer in vogue. We were all still "veiled," wearing the hijab for modesty. In fact, we had to twist Rasha's arm just to get her to agree to music. "Rasha, it's a wedding, not a funeral," my mother, sister Rula, and I exclaimed over and over again, in exasperation.

Most of the Western-style designer evening dresses we favored were sleeveless, short, or too low-cut, so we had our dresses made by a Cairo tailor to conform to our requirements for modesty. One evening, my mom, sisters, and I were at the tailor's for a fitting when my cellphone rang.

"Hi, Rana? This is Nicholas Negroponte."

The telephone line was scratchy, with lots of static that made it difficult to hear. I wanted to make sure I was hearing the name correctly. So, I asked, "Who is this?"

"This is Nicholas Negroponte. I am calling to make you an offer to join the MIT Media Lab as a postdoc."

It was the moment that reset my life: There was no question in my mind that I was going to accept this offer. When I began my PhD at Cambridge, I had no idea how those three and a half years were going to change me. I had been exposed to a new way of life and to the possibility of new horizons, and it was impossible to undo all that. Also, I had left too much unfinished business behind. What if I could build something that really helped autistic kids? What if this technology changed for the better the way human beings "connected" online? Wasn't that too important to walk away from?

When I got the offer to join the MIT Media Lab, I basically told Wael that I was going. I didn't flat out say, "Whether you like or not, I'm pursuing my dream." But I didn't ask for his permission, either. And that was a radical departure from how we had interacted up until this point. I had always sought his advice on just about everything and believed that his opinion was more valid than mine. For the first time, I felt confident to make up my own mind. I felt as if I had something important to contribute to the world and I wanted to see it through.

I have to admit, though, I was a bit queasy at the thought of how he might react to my leaving again, but I pushed these thoughts aside. I was bubbly and enthusiastic and didn't want to think about any of the potential negatives. This time, I told him (and myself) that it would be different. Although I still had to work out the specifics of my arrangement with MIT, I told him that Roz was fine with my working remotely most of the time, and visiting MIT only every few months. If there was any concern in his voice or expression, I completely missed it. He seemed supportive of my taking this job—well, at least he didn't object—and I took him at his word.

Looking back, I wish now that I had video footage of that encounter, to see if I missed some nuanced, subtle, fleeting expression of disapproval, disappointment, or even contempt on his face. It may have been a massive EQ failure on my part, but to me, Wael looked and sounded fine. But perhaps I didn't want to see any negative feelings. Maybe Wael was conflicted about not wanting to stand in my way yet still wanting me to be a "normal" wife.

I did feel guilty about leaving Jana. She was in nursery school in Cairo, and I didn't want to take her back and forth with me to Boston. When I left for Boston, I may have been more upset about my departure than she was! The tears flowed freely down my cheeks. She gave me a big hug and kiss goodbye, but she didn't beg me not to go.

If she had, it would have been very hard for me to leave her. But Jana was used to being cared for by others and adored my in-laws, who often looked after her when I was in Boston. I missed her terribly when I was away. We had grown so close when I was working on my PhD, and she was thrilled when I came home. But she was also doing very well in my absence, and that was a great relief to me.

I arrived in "the other Cambridge" on February 5, 2006, to begin my postdoc at the MIT Media Lab. Five days later, a fierce nor'easter dumped fifteen inches of snow in Boston. It was cold, stayed cold, and got colder, with temperatures dipping into bone-chilling digits. As bad as Cambridge, England, could be in the winter, I had landed in a place where the winters were even worse—how was that possible? So, the fact that I stayed, and kept coming back for more, is a testament to how much the place meant to me. It is not like any other computer lab in the world.

Founded in 1985, the MIT Media Lab was conceived by Negroponte as a tech incubator to prepare for "the convergence of media," the blending of computers, newspapers, television, and other modes of communication. Negroponte recognized that this blending would transform society. Unlike the conventional "computer" lab, the Media Lab prides itself on being "antidisciplinary," or, more accurately, interdisciplinary. Yes, everyone at the lab is adept at programming; that goes without saying. But computer science is not their only passion. The motley crew of musicians, neuroscientists, physicians, artists, designers, educators, and psychologists—even a professional magician studying the science of "wonder"—who make up the team were all here, like me, to bring their "crazy" ideas to fruition.

Despite the different disciplines all gathered under one roof, there was a unifying theme to the lab: We were creating technology to improve human life, and that was the glue that held this disparate group of misfits together.

Similar to the structure that is home to the computer lab at Cambridge University, the Weisner Building was boxy and hard-edged. But the similarities ended when I walked through the glass doors and into the main area. When I first looked around, I was a bit startled. It was pure chaos, with stuff strewn everywhere: piled on tables and plunked down in a big, open loftlike space. Unlike the Cambridge Lab, here there were no tidy, organized rows of desks or cubicles; rather, one group flowed into the next. Every space was a "maker space." Tinkering was encouraged, even expected. There were small offices tucked away against the wall for faculty, with doors that shut, and official "work stations" for the groups, but we often hung out on easy chairs or couches in the communal area. At the time, the lab was in the midst of expanding into a shiny, new glass extension being built around us, which added a bit to the overcrowding and chaos. (The new Media Lab building opened in 2010.)

In contrast to Cambridge, where students and faculty were buttoned up and "proper," here students showed up to work in sweatpants, and there were a few who I suspect didn't bother getting out of their pajamas. Nobody cared.

I confess that after three-plus years at the more formal Cambridge, and another year at the more sedate AUC, here I felt liberated to follow my own direction. No one was trying to pigeonhole me, as in "Oh, you're a computer specialist; stay in your lane." Instead, the lab allowed intellectual freedom in a way I had never experienced before.

The Media Lab is unique in many ways, but especially in terms of how it is funded. Most of its seventy-five-million-dollar annual operating budget comes from industry sponsorships, not from the government. These eighty-some "sponsors," or members, include some of the top companies in the world, tech companies like Google, Samsung, and Twitter, but also so-called non-tech companies (if there is such a thing anymore), like Twenty-First Century Fox, Deloitte, Estée Lauder, Benz Research and Development North American, and the Lego Group.

Private funding freed the lab from dependence on government grants, but that can have its downside. In September 2019, Joichi Ito, the lab's third director, resigned because he was accused of trying to cover up contributions to the lab from financier and alleged sex traf-ficker Jeffrey Epstein. All of this occurred years after I had left the lab.

Roz's group, Affective Computing, is tucked away on the ground floor, next to the Biomechatronics Lab, headed by Hugh Herr, a dashing double amputee from the knees down who sprinted around the lab wearing two biomechanical lower limbs designed by his group. I was awestruck to find myself a stone's throw away from the Lifelong Kindergarten Group, inspired by the late Seymour Papert, PhD, a renowned mathematician who proposed the radical notion that children should be taught programming skills in school. This was at a time when only an elite few adults knew programming. Pa-pert was co-inventor of the Logo programming language. I knew it well. It was the software I had used in elementary school to learn how to write my very first piece of code. It enabled the user to draw a Christmas tree on a computer, complete with twinkling lights. And it had started me on the road to computer science.

When I arrived in Roz's group, I was in awe of the different proj-ects under way there. The Affective Computing Group was working on iCalm, a wristband that tracked an individual's sympathetic ner-vous activity in real time to measure stress levels and sent the data to a laptop or mobile phone in the form of an accessible readout. This was Roz's baby; she always wore her iCalm wristband around the lab.

Hyungil Ahn, a Korean PhD student, had just build a prototype of a RoCo, a robotic computer that moved its monitor (its "head" and "neck") to interact with users in a playful way, aiming to improve their posture. Later, Hyungil and I collaborated on a project that combined my facial analysis technology with his interest in under-standing the difference between "liking" and "wanting." We ran a multi-month study, a play on the famous Pepsi-versus-Coke taste test. We asked volunteers to try different flavors of soft drinks. (It was a blind study, as in participants didn't know what flavor they were trying or the brand.) Each time they took a sip, we would quan-tify their response, moment by moment. Did their nose wrinkle and their head pull back? Well, that meant they didn't really like that flavor. Did they raise their eyebrows and lick their lips? Hmm, that meant they were intrigued by the flavor. This work marked the first time anyone was able to quantify, moment by moment, how people responded to new products. Not surprisingly, this research caught the attention of a few of the sponsors, like Procter and Gamble and Bank of America, who were curious about how this application could be used to test consumer experience as it unfolded over real time.

But perhaps the most fascinating project of all was Seth Raphael's research on "In Search of Wonder: Measuring Our Responses to the Miraculous." Seth was a magician whose claim to fame is combining technology and magic. For instance, he once asked me to think of an object, of any shape or size. Of course, I thought of the Egyptian pyramids. He then used his tablet to google something and, lo and behold, a picture of the pyramids popped up on the screen! A good old-fashioned magician, Seth has yet to reveal to me how he did that. Anyway, I'm sure he saw the expression of surprise (wonder!) on my face—and that was precisely what he was focused on.

If I was the buttoned-up, professional, no-nonsense member of our group, Seth was the complete opposite. He often showed up in crazy colored suits, green or red, and he couldn't have cared less what people thought. How freeing!

I felt increasingly at home at the MIT Media Lab. Certainly, I was more attuned to the people there than anywhere else I had ever studied or worked. The problem was that it was 5,400 miles away from my real home. Still, I was able to work out an arrangement that enabled me to be in Cairo most of time—I often joked that "I had the longest commute of all." Every fall and spring, the lab held "Sponsor Week," when the Fortune 500 companies that fund the lab are invited to see the students demo their projects. I had promised to be there for those events. In addition, I would move to Cambridge, Massachusetts, during the summer with Jana and work full-time at the lab. During the year, I would visit for two weeks every few months or so.

The flexible schedule had its advantages and disadvantages. I was straddling two worlds. Theoretically, I was clocking in more time at home than I was in Cambridge, but I felt like I was neither here nor there. When I was in Cairo, I was working Boston hours, trying to stay in touch with my group and always thinking about the next trip to the States. And when in Boston, I was worried about what was going on back home. I felt torn.

The Media Lab is halfway around the world from Cairo, but in truth, it is more than just geography that sets it apart. There is a completely different worldview at MIT: You didn't get brownie points by being obedient. Just the opposite: Being an intellectually defiant, disobedient misfit was cool. Challenging the norms was what the lab was all about, and that was exactly what I was doing.

The fact that I, a married Muslim woman with a child, leading a conservative lifestyle, had broken with the stereotype of what women like me were supposed to be made the lab a perfect fit. In my own way, I was as rebellious as any of them. Perhaps, considering the culture I was raised in, even more so. I had found my tribe!

Because I traveled so much between Cairo and Cambridge, Massachusetts, the juxtaposition of the two cultures became heightened in my mind. In Egypt, you get dinged for taking risks. It's discouraged in a conformist society. You don't want to call attention to yourself, you don't want to stand out. It could make you a target; it could cost you your career, your relationships, and in some instances, your life.

At the lab, being bold, thinking big, and taking risks were rewarded. And the bigger the risk, the better. It didn't even matter what came out of that risk taking; any risk taking was a success of sorts. Because when you take a risk, when you take a leap of faith, you build something new, and even if it fails, you learn from it.

As I "commuted" between Cairo and Cambridge, Massachusetts, every few months or so, I struggled to reconcile my two worlds. My family was already beginning to ask, "Why don't you just take a faculty position at AUC? After all, wasn't that your plan all along?"

Yes . . . and increasingly, no. People typically talk about having dreams and spending their life chasing them. They seldom talk about outgrowing their dreams. My dream since I was a freshman at AUC had been to become a faculty member there. But my experience in the United Kingdom and now my postdoc at MIT had opened my eyes to what could be, and how I could play a role in shaping our technology and our future.

This was a critical time in technology. We were rapidly moving into a new era of mobility that put powerful computing tools in the hands of individuals, and this opened the door to new ways of interacting with and learning about people. For me, there was no turning back. I wanted to be part of this new world. I wanted to be in the lab where it happened, and I wanted to be the one who steered this technology in the right direction—and I couldn't shut off that desire any longer.

In one respect, I fit in beautifully in a place filled with misfits and dreamers. Up until then, I had been the one with the "crazy" ideas. But at this lab, there was some steep competition for crazy. When I joined the lab, being a woman wearing a hijab—the *only* woman wearing a hijab—made me *very* different, practically a novelty act. People didn't really understand my dress. In Cambridge, UK, almost everyone I met had traveled around the world, and many of my colleagues had been to Egypt, or at least another Arabic-speaking country. But here in Cambridge, Massachusetts, many people had never traveled outside the United States, and even when they had, very few had gone to Egypt. Their perception of Egyptians and Muslims was heavily shaped by the media; they knew very little about us. I don't think they expected a religious Muslim woman to be a scientist, let alone a scientist in a lab known for its maverick approach.

I was always treated with respect, even generosity, but sometimes I felt as if I were regarded as an exotic creature or an "extraordinary alien," to use the term U.S. Immigration did on my work visa. (That label always cracks me up; I picture a green one-eyed monster like the one in *Monsters, Inc.*) And as the "extraordinary alien" in the hijab, I would get asked some extraordinary questions: "Do you wear the scarf because of medical reasons?" or "Is your hair wet, and you only wear it until your hair dries?" or "Do you also wear your hijab at home?" People didn't know what to make of my hijab or me. And they were equally mystified that I never drank alcohol, or that, in this era of sexual liberation, with people hooking up with one another all the time, I had dated only one man, ever—my husband.

Still, in an environment where outliers were appreciated, in my own way, I fit right in. There had never been anybody like me at the lab. And despite cultural differences—and there were many—we all worked alongside one another and found common ground, and connection, in the technology we were trying to build.

PART III

Straddling Two Worlds

13

The Other Cambridge

Once I became part of the Media Lab, I was filled with evangelical zeal to bring the incredible technology and innovative spirit of the lab back to my native land. I knew that Suzanne Mubarak, wife of then-president Hosni Mubarak, was in the process of planning the Suzanne Mubarak Family Garden, forty-two acres of "hands-on" centers in Heliopolis that today provide learning and fun experiences for children and adults. I immediately saw a potential partnership with some of the Media Lab's projects, such as its Lifelong Kindergarten Group. In the past, the Media Lab had helped launch the Children's Art Museum and Park (CAMP) in Tokyo, a workshop center that uses music, technology, and art to foster creativity. Something similar would be perfect for Cairo.

Frank Moss, then-director of the Media Lab, gave the plan a thumbs-up, and I leveraged my connections in Cairo to get an introduction to Mrs. Mubarak. I then approached her with the idea of the Media Lab partnering with the Family Garden. A sociologist by training, Mrs. Mubarak was enthusiastic and asked me to come speak to the planning group developing the project.

The meeting was held at one of the Mubaraks' Cairo homes, in Heliopolis. (The family had eight official residences.) I arrived at the front gate and was escorted into a conference room. About half a dozen guests were already sitting around a large oblong table. I noticed that I was the youngest person in the room by a long shot. I was surprised that no one made a move to suggest that I sit down and join them. Instead, one of the men sitting at the table, Zahi Hawass, then the former minister of state for antiquities affairs, looked up at me and gestured over to a table at the other end of the room, where coffee, soda, and water were set up. He didn't exactly say, "Water with no ice, please," but the implicit gesture was definitely "There are the drinks. Go do your work."

Actually, I was more amused than insulted. I didn't have time to correct him, because a second later, Mrs. Mubarak entered the room and came over to shake my hand. She sat next to Hawass and announced, "This is the MIT expert we invited today as our guest of honor."

I didn't need to be a certified face decoder to figure out Hawass's expression: complete surprise, even astonishment. That same look was pretty much shared by everyone around the table. I'm sure that when they first heard that the speaker was an expert from MIT who would be talking about bringing the latest technology to Egypt, they did not picture anyone who looked like me. I was female, and too young and "religious," which of course meant I couldn't possibly be a successful scientist, let alone at one of the top schools in the world. Without uttering a word, I had made an important point.

I stood in front of the group and began my talk. I had given this speech a great deal of thought. Here were the top leaders of Egypt's creative and intellectual class, and I knew that this was an opportunity to make a difference. I started out with a story, or perhaps more of a cautionary tale.

"I stay in touch with many of the students that I taught when I was at AUC, and several of them are working on their senior thesis. So, I tell them about all the hot new ideas from MIT, and recommend that they consider doing something like this for a project," I said. "The most frequent comment I get is 'But, Doctor, this idea is *so risky;* it's never been done before.' *Yes!* That is the whole point. In contrast to Egypt's risk-averse culture, at MIT you can't suggest an idea that's been done before. The key is to be creative. You have to explore something new, even if it fails, and that's okay."

By now, I had everybody's attention. A few were nodding in approval, and some clearly did not approve. I kept going.

"Whereas it was okay twenty years ago to have a conformist society, we can't get away with it anymore. Today's economic growth is all about talented, creative individuals who are fearless. It's about innovation and being different. And I believe that if we want to be part of this economic growth, part of this rising creative class, we have to start with our children today and foster this passion for learning, innovation, and creativity."

In the end, the group voted to participate with MIT, but we hit a snag when it came to an agreement on who would fund it. MIT didn't want to pick up the tab, and neither did Mrs. Mubarak. And so, much to my disappointment, the talks eventually broke down.

LESSONS IN LIFE

The more time I spent at MIT, the greater the contrast with my own culture. In particular, Roz, my mentor, adviser, and role model, was instrumental in shaping my view of this new world. And opening my eyes to new possibilities. We couldn't have come from more different backgrounds: Roz was raised in an atheist household and converted to Christianity as an adult. Even beyond our professional relationship, I learned a great deal about life by observing how she conducted hers.

Although Roz was an internationally recognized scientist and super-successful career woman, she was married, with three school-age sons, and she placed tremendous importance on family. When I was in town, she would always invite me over for dinner. I will never forget the first time I visited her home in a Boston suburb. When she introduced me to her husband, Len, an engineering manager at a publicly traded Boston-based company, she added that he was on "cooking duty" that night.

I was puzzled. I asked, "You mean *he's* cooking dinner?"

"Of course," Roz replied, and then explained that because they both had full-time jobs, pretty intense ones, too, they alternated who would cook dinner every night. Roz was assigned Mondays, Wednesdays, and Fridays—she was super organized and planned everything down to the smallest detail the night before—and Len was the chef on Tuesdays, Thursdays, and Saturdays. Sunday was a free day, which I guess meant they ate out or shared the cooking duties.

As it turned out, Len (who was of Italian descent) had not only made a terrific lasagna (his grandmother's recipe) and set the table, but even served the food. When dinner was over, we all helped clear away the dishes. Perhaps the biggest shock of the evening was when I saw Roz's three young boys dutifully carry their plates to the kitchen. In my grandmother's house, when all the cousins were around, the minute the meal was over, the men would vanish to the patio for tea, and the boys would run into the garden to play. That left the women and girls with piles of dirty dishes, food to wrap up and put away, and dessert to prepare. Traditional gender roles were so fixed in my brain that it was surprising, even uncomfortable, to see Len clear the table and load the dishwasher. I kept offering to help, but he firmly rejected my offer. "No! I load the dishwasher; I like it organized *my* way." Wow! That was a real eye-opener: men can do this job!

I grew up in a pretty modern household for the Middle East, in that my mother worked a full-time job. But even though she was "allowed" to work, she was still solely responsible for feeding us *every single freaking day of the week*. And my extended family was no exception; it was the same in my aunt's and uncle's households. I had never seen my dad (or any guy, for that matter) load a dishwasher. As progressive as Wael was, he really didn't do much around the house, either.

I figured Roz had found herself the best husband ever. Surely, there were no other men on the planet who came close to Len. But as I got to understand U.S. culture a bit more, I saw that it was a lot more common to find spouses who participated in the household; maybe not always equally, but American men were expected to do *something*. Still, Roz and Len had raised the bar of what relationships should look like; they were the first example of truly equal partners I had ever seen. I was envious of their relationship. While my marriage was modern by Middle Eastern standards—unlike my mother I could talk about my work at home and I could travel on my own—it wasn't like Roz's marriage. My "enlightened" husband allowed me to come and go, and pursue my career. But that was very different from having the innate freedom to manage your own life, or have the true collaborative relationship that Roz and Len had.

I also admired the way Roz found time to be there for her boys. One of her boys fenced, and she would always take him to fencing practice and tournaments. She would bring her laptop along and work while waiting for him, which is something I do now when I take my kids to their various activities.

Roz was also incredibly organized. Every block of time on her calendar was accounted for, and that is how I set up my calendar now. She even made time to exercise or jog every morning, and Len biked to work. Both inspired me to prioritize exercise in my life.

I come from a part of the world where many countries are intolerant of other religions. Indeed, if you're not Muslim, you are not just misguided, but headed straight to hell. Now, I had attended British schools in Kuwait and the UAE, and I had had Christian friends at AUC, but for most of my young life, I was steeped in Muslim culture. As I became close to Roz and Len, and got a chance to observe their strong core values and high integrity at work, I began to question how people like them, who lived exemplary lives, would be headed for hell while someone who was Muslim and who was corrupt or treated people badly would get an automatic pass to heaven just because he prayed five times a day. That made no sense to me at all.

While I still believed deeply in the guiding principles of my faith—kindness, generosity, compassion, service, hard work, and living the best life you can—that day at Roz and Len's made me question my assumptions about the practice of religion.

14

Demo or Die

spent the summer of 2006 working at the lab, continuing to build on the core facial analysis engine I had developed at the other Cambridge. I called the new and updated algorithm "FaceSense," essentially version 2.0 of the Mind Reader. For now, FaceSense was limited to a laptop or desktop, with a fist-sized Logitech webcam perched on top. But our goal was to have iSET, a wearable Google Glass–type device for autistic users that could run FaceSense, up and running by Sponsor Week in the spring of 2007.

Our crazy idea now had a home. But just as the National Science Foundation reviewers had predicted, iSET turned out to be a nearly impossible build. Indeed, we had to tackle one seemingly insurmountable challenge after another. Remember, our design required a camera tiny enough to attach to a pair of glasses, a computer small enough to wear or carry around but powerful enough to run the FaceSense program, and ear buds that would deliver messages to the user via Bluetooth technology ("Friend is happy, Friend is confused"). And because none of this technology existed in the right form, we had to invent it as we went along.

We hit the ground running, but ran smack into a wall. First, even the most state-of-the-art cameras designed for computers, webcams, were still huge for our purposes, and we couldn't exactly put a clunky Logitech webcam on a pair of glasses. We needed a camera that was small, but with strong enough resolution to detect subtle facial expressions.

We were stumped; where could we find one of those? Then it hit us: Who uses miniature cameras? Spies! This revelation opened up a whole new world I didn't even know existed. What really surprised me was that, as far back as World War I, watches, cigarette lighters, and all manner of objects had been embedded with secret cameras. Tiny cameras were even strapped onto carrier pigeons to send messages to the front. And now civilians were using these tools to spy on their nannies, nail unfaithful spouses, and weed out insurance fraud and industrial espionage.

I didn't like the idea of using technology to spy on people, but I felt that our objectives were pure. Our goal, after all, was to enable kids on the autism spectrum to read the nonverbal signals of others while engaging in a conversation—nothing sinister there. So, we ordered a few spycams off the Internet to see if they worked. The first few did not. The images were blurry, and we were very disappointed. But eventually we found the perfect spycam. It cost only eighty dollars, had sharp resolution, and fit on a pair of glasses.

The next hurdle was finding the right tablet. Obviously, a smartphone would have been ideal, but one didn't exist yet. This was even before the Kindle or the iPad. We decided to use an ASUS tablet, advanced for its time but not ideal; it was heavy and hard to program.

By early winter, after months of cobbling and tinkering, we had gathered all the pieces. Each one did its job independently, but would they work together? In Egypt we have an expression, *tala3 komash*, which literally means, "It produced the fabric!" It's a throwback to the days when Egypt was a leader in the cotton industry. A textile factory could have more than half a dozen different machines work-ing together to spin the cotton into fabric, but if there was a glitch in just one machine, the process would grind to a halt. Egyptians use it today to describe a successful effort.

The moment of truth: the set of glasses with the spy camera at-tached to it. A wire connecting the camera to the tablet ran along one of the glasses' temples. (It wasn't the most fashionable set of specs.) I put on the glasses, stuck the tablet into a shoulder bag, and slid the Bluetooth earpieces onto my ears. I then faced Alea Teeters, a master's student working with me; Alea was extremely creative and good with tinkering and hardware. If the system worked, the camera would detect her face and send that image to the tablet, which was running FaceSense. FaceSense would then analyze it, detect an emo-tion, and send the word for that emotion to the Bluetooth earpiece in my ears.

Alea smiled.

I waited, counting the seconds. Then I heard, "Smiling. Probabil-ity 90 percent."

It worked. *Tala3 komash!*

Now I needed to make sure it worked well no matter who had the glasses on or who was expressing the emotion.

In March 2007, on the eve of Sponsor Week, I stayed late at the lab to prepare, as did 99 percent of the other students and postdocs. (After all, our mantra during Sponsor Week was "Demo or die!") Tomorrow, we would be demo'ing two projects.

I wasn't worried about FaceSense: I had perfected that algorithm since my Cambridge days. It operated reliably and used a standard computer and webcam. But the wearable device, iSET, wasn't above throwing a last-minute temper tantrum. For months, I had tweaked it in order to have a working version to show at Sponsor Week, when hundreds of sponsors would flood the lab. Alea and another master's student, Miri Madsen (now an MD), were helping. Miri, who was good with visualizations, was responsible for mapping the output of the program into something visual that people could understand.

Sponsor Week didn't really last a full week—it was more like a nonstop three days—but it was very intense. By eight A.M., I had my laptop set up for FaceSense and attached to a big screen with a camera mounted on top. The iSET glasses rested on a table.

Sponsors started to arrive in waves. They made faces at my laptop and saw their facial expressions analyzed by the software. They tried on the iSET headset and interacted with the tablet; miraculously, it continued to work. We were a hit.

A top executive at Gillette wanted to talk to us about possible applications of our technology to test product experience—specifically, he wanted to understand what men (and women) *felt like* when shaving. Could we detect when nicks happened and quantify that?

A team of producers from Fox wanted FaceSense to observe the reaction of a test audience to the network's upcoming fall lineup of new shows.

Procter and Gamble wanted FaceSense to test how people reacted to different shower gel scents.

Toyota wanted our software to detect distracted or drowsy drivers.

Then some executives from Logitech, the webcam people, came along. I showed off my spycam iSET rig. One of the executives asked about the specifications of our spycam and said, "Oh! Our latest optical sensor totally beats that; we'll ship over a couple of our latest cameras." That was the true magic of the MIT Media Lab, where innovation happens at the intersection of different disciplines. A place where there is continuous cross-pollination of ideas across industries, geographies, and disciplines. Our sponsors didn't just take our knowhow and say, license it. They often helped shape our idea, applying it in ways that we, as scientists, hadn't thought of.

Sponsor Week was grueling work. By day two, I had lost my voice and had to speak in a whisper. But I was on an adrenaline rush. More sponsors, more ideas—I found it exhilarating that a technology that had been developed for autism had so many varied and imaginative potential uses, things that Roz and I might never have dreamed up on our own.

Part of the agreement between the lab and its sponsors was that sponsors enjoyed access to our work on a password-protected online repository. Students and faculty could post any program online they thought would be of interest to the sponsors. I put FaceSense online in 2007, and by 2008, it exploded, quickly becoming the most downloaded program at the lab at the time.

Working with sponsors gave me a sense of the breadth and scope of our new technology. But the most memorable experiences I had during my years at the MIT Media Lab were the times I spent with teenagers on the autism spectrum. With a working prototype in hand, we had reapplied to the National Science Foundation, this time for a much larger grant, which we received. As part of the grant, we partnered with a progressive school for autistic children, the Cove Center in Providence, Rhode Island. The Cove Center is part of the Groden Network, a leading provider of services to the autistic community and to other children and adults in the state with developmental disabilities. Its head of research, Dr. Matthew Goodwin, a passionate young clinical psychologist, was a big believer in the ability of technology to transform the field.

I was adamant that we take a design-thinking approach, which meant involving the actual users—in this case, the kids and their families—in helping design the system. These were mostly high-functioning teens (and adults) who were out in the world, interacting with peers. The teens suffered deeply because of their inability to understand nonverbal cues. Like most teens, they wanted to spend time with their peers, date, go to parties, and do everything other kids did, but because of their social awkwardness, they were excluded and even targeted by bullies. Because they couldn't differentiate between a polite smile and a flirty "I'd like to get to know you" smile, or identify an eye roll that meant "Stop talking about that topic; it's boring"—this behavior in people on the spectrum is called "monologuing"—they were at a severe disadvantage. They didn't understand sarcasm at all (and teenagers tend to be *very* sarcastic), and their lack of social skills kept them on the periphery of mainstream life, unable to make or keep friends or even hold down a job.

These teens' lack of emotional intelligence also strained their parent-child relationship. At one point, we invited eight pairs of mothers and sons to try out the technology and talk about some of their challenges in interacting with each other. I remember one mother and son in particular: He thought his mother was always angry at him, and he didn't understand that fluctuations in voice can have many different meanings. To him, anything that seemed "loud" was anger. The iSET showed him that his mother was not always angry, and helped him better understand that there were subtle emotional states that were not apparent to him. We learned a great deal by working with the teens and young adults, and some of them had good design suggestions.

But perhaps the most powerful moment came about six months into the project. By then, I had been to the Cove Center many times, and I knew the kids pretty well. None of the boys there ever made direct eye contact with me (or with anyone else, for that matter). That is one of the core problems with autism: Individuals on the autism spectrum find faces so overwhelming that they avoid them altogether. But for them to become more adept at reading emotions, we had to get them first to *look* at a face. And because they often used the iSET tablet as a barrier to direct face contact, we devised a game with it: They got points every time they looked directly at another person's face. After months of trying this with the kids, I was roaming around school when one of the high-functioning teenagers stopped, lowered his tablet, and looked me straight in the eye. Our gaze locked for a few seconds. In this one moment, this young man and I were connected on a basic human level. It was authentic, vulnerable, and powerful all at the same time. It had all been made possible by the technology that I had built. It was once again a reminder that my work in the tech space was primarily about connecting people together at a deeper level.

Working with these young men reinforced my belief in the importance of reading emotion and that the same "emotion-blind" technology that robs us of this vital skill can be reimagined, with Emotion AI, to enhance our understanding of one another.

By my second year at the Media Lab, in many ways I began to feel more at home in Cambridge, Massachusetts, than I did in Cairo. Back in Cairo, no one seemed to know what to make of me, a hijabi mom who disappeared for a week every month or so to work at some school in Boston. I found it odd that my friends and family never asked me about my work at MIT. I don't know why they didn't; maybe I made everyone too uncomfortable. I was definitely not behaving like a nice Egyptian wife!

I loved the energy and entrepreneurial spirit of the Media Lab. I had done so well my first year in terms of my work and bringing in funding that I had been promoted to research scientist. I decided that I wanted to apply for a tenured faculty position. That would require a far greater commitment in terms of time at the lab. I still worked a good deal of the time remotely, and that would have to change. I raised the possibility with Wael of us moving to Boston so that I could be on the faculty, and he told me that this was not the right time for him to move. The following year, before the deadline to apply for a faculty position had expired, I asked him again if he would consider moving to Boston, and again, he said not now. So I didn't apply.

Wael believed he could have a bigger impact on society by staying in the Middle East, where he was already well-known and respected, continuing to expand his company and encourage investment throughout the region. Our game plan, after all, was for me to apply for a faculty position with AUC, which is a wonderful school, but unlike the MIT Media Lab, the focus there is on teaching, not research. At the time, it did not offer a PhD program. Given my Cambridge PhD,

I was a strong candidate for a faculty position at AUC, yet I couldn't bring myself to apply for one.

I dropped the idea of becoming part of the full-time faculty at MIT but continued my work with Roz as a research scientist, which meant I spent a lot of time on airplanes. On one of my sixteen-hour flights from Cairo to Boston with a stopover in London, I read *The Namesake*, the Pulitzer Prize–winning novel by Jhumpa Lahiri. The story revolves around Ashoke and Ashima Ganguli, Bengali immigrants from India, and their adjusting to life in the United States. In the late 1960s, Ashoke brings his new bride, Ashima, to Boston to accept a fellowship in engineering at MIT, leaving behind a large and close-knit family. Over time, their children become more and more Americanized, torn between their new country and the traditions of their old country.

Something about this story hit a nerve. I began to sob. Deep down inside, I knew I was at an inflection point. I was embarking on a journey that would take me away from Egypt, and perhaps my family. I was becoming, as a friend put it, "Americanized." I was challenging the status quo, and rethinking my religion and the roles of men and women. I saw firsthand how academic freedom and the right to self-expression enabled scientists to do their best work. I wanted to spend the rest of my career surrounded by fearless, bold, imaginative people who weren't frozen in place for fear of making a mistake. These thoughts scared me because they placed me at odds with my culture, and I couldn't see my way forward. At the time, it was too frightening to think about where that road led.

That spring I attended an international conference on autism, where I bumped into Ofer Golan, the Israeli graduate student I'd met back at Cambridge University who worked with Simon Baron-Cohen. It had been five years since we had seen each other, and I was astonished— even jealous—to hear that during that time, Ofer had managed to have two children. I was suddenly struck by the fact that Jana was already five years old and was still an only child. Although Wael and I had every intention of having a second child, our lives were so hectic, each of us preoccupied with work and me trying to spend as much time as possible with Jana when I was home, that we seemed to have pushed aside the thought of having another baby. I was turning thirty that summer and worried that I wouldn't be able to get pregnant as easily as I had the first time. I was close to my sisters and loved having siblings. I wanted Jana to have the same experience. Of course, to make a baby you need to have sex, and I couldn't remember the last time Wael and I had been intimate, but I shrugged that off. We weren't just a two-career couple, trying to juggle family life and work, we were a two-career couple where one partner worked in a different hemisphere. Our lives were complicated. But still, I felt a longing for another child.

When I returned home, I told Wael it was time we gave Jana a sibling, and he agreed. About a month later, Wael and I went out to dinner at a seafood restaurant. When we got back home, I threw up. I was thrilled. I knew I was pregnant again. As with my first pregnancy, I was nauseated a good deal of the time, but I worked around it. I carried crackers in my bag, sipped ginger tea to calm my stomach, and ran to the bathroom when I needed to throw up.

Wael and I agreed that I should have the baby in Boston, rather than Cairo. I was becoming increasingly involved in my work at the lab and felt it would be easier for me to stay in one place, especially in the later months of pregnancy. I found a midwife through the MIT network, purchased another birthing ball, signed up for yoga classes for expectant mothers, and was incredibly happy. This was a very different time in my life from when I was pregnant in England. I no longer had anything to prove. I had earned my PhD, my research was going incredibly well, and Roz and I were getting very positive feedback from the lab's sponsors. I was excited at the prospect of having a second child.

THE TIPPING POINT

By summer 2008, more than twenty sponsors were interested in pursuing projects with us. But MIT code is not "commercial-grade." It shows proof of concept—yes, this can be done—but it is not reliable enough to go to market with and make it scalable. In other words, a company couldn't just take our code and work it into their product. I had to be there to get it up and running.

We continued to get calls long after Sponsor Week, but we didn't have enough research students to respond to all the requests. So, in late fall, armed with our list of interested sponsors, Roz and I barged into the office of the MIT Media Lab director, Frank Moss, and pleaded, "We're inundated. We need more research fellows."

Prior to assuming the directorship of the Media Lab, Moss was a successful serial entrepreneur. To him, the answer to our problem was obvious.

Frank looked at both of us and said, "Nope. You don't need more students. You need to spin out." In plain English: We needed to form our own company.

My knee-jerk reaction was "But I'm an *academic*." Starting a company—in the United States, no less—was definitely not in "the Plan." Oh, by now I was six months pregnant, another reason I was reluctant even to consider such a radical and daunting direction.

More to the point, I believed that I had no head for business. The world of finance and deals was Wael's domain. Just that past September, he and I had been on a mini-vacation when the U.S. stock market crashed. The Dow Jones Industrial Average dropped 777 points in one day, the biggest one-day drop in its history up to that point. Worried, and keenly aware of the ramifications that a sinking U.S. economy would have on business everywhere around the globe, including Cairo, Wael wouldn't leave the hotel room; he was riveted to CNN. What did I do? I took out my laptop and caught up on some work, completely oblivious to what was going on. That's how little I thought about business, the markets, and money.

I did, however, care about seeing our technology in the hands of the people who could benefit from it. The more time Frank, Roz, and I spent talking about the pros and cons of forming a company, the more I began to see that if we wanted to develop our research to its fullest potential, spinning out was our only option. If we stayed in the Media Lab, we would be constrained in doing everything we wanted by lack of both money and human resources. If we formed a company, though, and raised money, we would have a unique opportunity to bring our creations to the world in a way that could change how people interacted with technology and one another. We had a chance to make a positive impact on the lives of people everywhere. For me, that was the tipping point.

Since the baby was due in February and I didn't want to travel in my third trimester, Jana lived with me in Cambridge that fall. Wael and I decided to enroll Jana at the British School in Boston. We agreed that Wael would stay in Cairo for now but plan to return closer to my delivery date. Every morning, I would walk Jana to the school bus and get home in time to pick her up at the bus stop in the late afternoon. We shared a small house near school with a PhD student, and in mid-January, my mother and Wael came to join us.

It was wonderful having Jana around full-time with me again. Since I had been commuting to MIT, I missed those days in Cambridge, England, when we were able to spend so much time together. One weekend, I took her to the Boston Museum of Fine Arts; visiting a museum in Boston is a much more engaging and interactive experience than the typical museum in Cairo. Jana loved it. We each used an audio guide with headphones provided by the museum and strolled around the vast exhibit of musical instruments. We listened to musical snippets demonstrating the different instruments on display—violins, pianos, flutes, and some more exotic ones, too. And then Jana stopped dead in her tracks as she stood in front of a display of harps, listening to the strains of harp music on her audio guide. She was mesmerized. "Mummy, I want to play this instrument!"

The harp in front of us was huge, a full four feet tall, taller than Jana at the time. I had no idea where I could even find a harp! I tried persuading her to take up the flute or recorder or clarinet or something more manageable. But she had her heart set on studying the harp. And so I found a harp class two T stops from MIT. It's the kind of thing that my mother would have done. She went out of her way to make it possible for us to pursue our interests; I was following in her tradition.

I was adamant about having a natural childbirth again. I insisted that the bed in the delivery room be moved against the wall to make my point (I had no intention of lying down until I absolutely had to), so I had room to pace and bounce on the birthing ball. When the contractions slowed down, I tried to speed things up by walking briskly around the hospital corridor wearing headphones playing "Last Summer" by Abba on repeat. Wishful thinking—it was snowing outside and freezing cold.

Adam was born at eleven A.M., February 4, 2009, in Cambridge, Massachusetts. Wael and I had considered several names for our son, but when we took our first look at him, I said, "He's definitely an Adam!" It was a visceral response. He just looked like an Adam! As he lay on my chest for the first time right after the delivery, I recited a prayer, "May he turn into a kind and compassionate young man." In this regard, my prayers have been answered.

The day after I delivered Adam, Roz visited me in the hospital. With me propped up by pillows in my hospital bed, we edited a proposal we were submitting to the National Science Foundation Small Business Innovation Research (SBIR), due the next day, to continue our work on autism. We were seeking funding to help us bootstrap the company.

Wael had to return to Cairo to run his business, but my mother stayed with us until March to help out with Adam and Jana. I didn't have a car, so we had to brave the brutal Boston winters on foot. I'd bundle Adam up, stick him into a BabyBjörn carrier, and walk in the bitter cold to Whole Foods after putting Jana on the bus to school. I was petrified that Adam would freeze to death, and I stopped often to check that he was still breathing. Today, of all of us in the family, Adam is the only one that is impervious to the cold!

I was back at the Lab shortly afterward. Planning for the new company was consuming most of my time. Unlike Jana, Adam was a colicky baby with frequent night awakenings. I was tired—sleep deprived—but I kept going. Before my mom went back to the UAE, she filled the freezer with marinated chicken breasts so that I could just fry them up for dinner. One evening, I threw a few chicken breasts into a frying pan, turned on the stove, and then went into the living room to check on my email. I lost track of time, and the next thing I knew, the stove was on fire, and it was spreading throughout the kitchen. I panicked. I forgot whether I was supposed to smother the fire, douse it with water, or just leave the house. I froze for a minute, and then scooped up Adam, grabbed Jana, and fled downstairs to our housemate, who called the fire department. We all ran outside to safety. Within minutes, four fire trucks pulled up, and fire fighters dashed into the house to put out the fire. The fire had been contained to the kitchen, but it was quite a mess. It was weeks before the kitchen was usable. I was living with a toaster oven and microwave in the living room. I felt humiliated by the incident, and apologized endlessly to my landlord and housemate, who were incredibly kind about it.

This story could have had a different, tragic ending. We all emerged unscathed from the fire, and for that, I am eternally grateful. I learned an important lesson from the experience: that I needed to bring more mindfulness and balance into my life. I had been working nonstop, and clearly, I was exhausted. But as I have learned, the realization alone is not enough to implement behavior change. I have managed not to burn down another kitchen, but the truth is, when you launch a start-up, it can become an obsession. It's always top of the mind, and today's 24/7 connected technology makes it difficult to unplug. I have a great deal of trouble switching from work mode to life mode and it has cost me, at times, dearly.

On April 14, 2009, Roz and I incorporated Affectiva.

Roz and I decided to invest our own money to get the company off the ground. But we knew that soon we would have to reach out to the investor community, many of whom were veterans of the tech industry. None of this came easily to us. We were complete novices, and despite the enthusiasm among our sponsors, in the world of tech we were swimming against the tide. We were still up against the antiquated "emotion is irrational" thinking that permeated the industry. Most of our potential investors were, to paraphrase Baron-Cohen, tilted more toward the "systemizer" end of the autism scale than the "empathizer" end, which is to say, they were not particularly comfortable with emotion, or "feelings."

As women cofounders, Roz and I were already anomalies in the male-dominated world of tech start-ups. So, we made a strategic decision to avoid using the word *emotion*, just as Roz had done a decade earlier when she coined the term *affective computing*. That is how we came up with the name Affectiva, an innocuous variation on *affective computing*.

In the beginning, we didn't exactly know what kind of company we had in mind. Given the interests of our diverse group of sponsors, we could have veered off in many different directions. We knew that there was one thing we absolutely had to establish: our core values. This was especially important for a company like ours, which deals with highly personal data about individuals' deepest feelings.

One early spring day, the newly formed Affectiva team—our two first employees, Jocelyn Scheirer and Oliver Wilder-Smith, and Roz and I—met at Roz's home to talk about our strategy and direction.

It was an unusually warm day; I remember the sunlight streaming through Roz's large windows as we met around her kitchen island.

We reviewed all the potential uses of our technology: marketing, education, mental health, automotive, autism, security, and surveillance. And we agreed on a set of core values that would guide our decision making with regard to the industries we'd entertain and those we'd turn down. We wanted Affectiva to be *the* company people trusted with their emotion data. And we knew that in order for the company to be successful, people needed to feel confident in the integrity of our approach and the science. Unless people fully trusted us, we wouldn't be able to get the data that would enable us to help them. We knew that the moment we breached that trust, we would lose the very people we wanted to help.

We therefore decided that we would deal only with companies that agreed to our "opt-in" terms, meaning people had to be informed that their emotion data was being collected; most important, they had to consent to it. It was also essential that people have an option to opt out whenever they wanted. That meant we would not work with surveillance or security companies. Down the road, this did limit our options, but it was the right decision.

We rented space at One Moody Street—yes, that really was the name—in Waltham, a suburb thirty minutes from MIT. It was a run-down, rather drab office space—the wooden floors squeaked when you walked on them—but it was the perfect home for our young, scrappy team of four.

I couldn't bring myself to break the news to my parents that I was now working part-time at the Media Lab so that I could focus on my start-up, Affectiva. (Remember, my father didn't exactly have a high opinion of start-ups.) I kept the company a secret from them. Wael knew, but for two years my parents thought that I was still at the Media Lab full-time. I didn't want to deal with their attention and concerns at the same time that I was trying to launch the company, reach out to investors, and care for an infant—who I was still breastfeeding and who traveled back and forth with me. My parents believed that an academic appointment at a prestigious university was the pinnacle of success one could achieve. I didn't have the heart, or perhaps the courage, to tell them that I had walked away from that to pursue my own dream.

So much was going on in my Boston life that I couldn't talk about in Cairo. Roz and I were moving very quickly, making inroads in ensuring that Affectiva was a bona fide business. Well connected to the Boston business community, Frank Moss was generous with both his time and contacts. He put us in touch with Andy Palmer, another serial entrepreneur, one who specialized in working with "mission-driven" companies like Affectiva. Our goal was to be a sustainable, thriving business, but also to do social good, especially around helping individuals who struggled with communication and regulating their emotions. Andy loved our dual mission, and became our first adviser.

Most start-ups (90 percent) go belly-up within a year or two. It's a daunting figure, one of the reasons I didn't want to tell my parents, "Hey, I just launched a start-up." Yes, Roz and I did have some advantages over the typical start-up. We already knew that there was a market for our product; we had a long list of sponsors who had expressed interest in using our software for all sorts of things. Still, the odds against any new company are overwhelming, especially for people who have never run one. Although Roz had raised millions of dollars for the Media Lab, and I had raised more than a million dollars myself, we were not business experts. Neither of us had ever dealt with the nuts and bolts of starting a company. Fortunately, we had access to MIT's Venture Mentoring Service (VMS), an amazing network of mentors, as well as other MIT spin-offs.

At VMS, we were assigned a team of eleven mentors who spanned a wide range of expertise. Some were venture capitalists, some had backgrounds in finance, law, intellectual property, marketing, or manufacturing. We could email any of them at any time with questions, no matter how mundane or stupid. One time, both Roz and I were thrown off by an email from an investor who was courting us (and who ended up being our first investor), asking for our "BS." The only BS that Roz and I knew of didn't make any sense in this context. We shot an email to one of our VMS mentors, asking what this meant. Our mentor must have laughed his head off.

Oh, our *balance sheet*! It was going to be a *steep* learning curve!

Once a week, we would meet with our mentors and pitch to them as if they were a team of investors, and in return, they'd tear our pitch apart. From their relentless criticism, we grew thicker skin, good preparation for dealing with a future of pitching to real investors. This went on for months until, one fall day, we had our act together. We were able to answer whatever questions they threw at us, our pitch looked professional, and we could recite in our sleep the vast potential of our technology. Finally, the VMS team looked at us and said, "Okay, you're ready. Go out and pitch this."

Next stop, Silicon Valley . . .

15

Spinning Out

The babysitter woke up sick and canceled on me at the last minute, but the morning meeting was too important for us to postpone. I flashed a smile at the receptionist and asked if she could keep an eye on Adam, who was all of nine months and utterly adorable as he napped in his portable car seat. "He's super well behaved," I assured her, handing her a milk bottle as I placed Adam on the floor next to her desk. The blond, blue-eyed young woman looked stunned, but grinned agreeably and nodded yes.

Roz and I raced into the conference room where the meeting was to be held, just in time to set everything up. A minute later, the sharply dressed fortysomething partner of a leading Silicon Valley venture capital (VC) firm joined us. We were seated at a giant rectangular walnut table in a wood-paneled room with a glass interior wall, with our Affectiva business cards in hand and our pitch deck, loaded with a PowerPoint presentation describing our new company, ready to go. Roz and I were wearing our "pitching" uniforms: drab gray (Roz) and brown (me) man-tailored pantsuits. No frills, no bright colors, nothing too feminine. I wore a matching brown hijab.

The partner shook our hands and then took his place in an uphol-stered armchair at the head of the table. He smiled and gestured for us to begin. But it wasn't a genuine, "glad to see you" smile, with his lips pulled up evenly on both sides; it was more of an asymmetric lip corner pull that was closer to a smirk. If this had been a scene in a comic book, I'm pretty sure the thought bubble over his head would have read, "Really, I've got to sit here for an hour and listen to these *women*? Who booked this for me?"

"You know," he said casually, "one of the first things VCs do is fire the founders."

Roz's eyes widened. I'm sure mine did, too—action unit 5, to be precise: a telltale expression displaying both surprise and fear.

He observed our reactions. "Just kidding!"

But was he, really? Tales abound of inexperienced founders being booted from their start-ups in favor of experienced guys with track records. Venture capitalists are notorious for investing in what seems safe and familiar, whether it's the team, the technology, or the idea. But in the fall of 2009, when Roz and I did the Silicon Valley VC tour to raise money for our start-up, we were anything but "safe" and "familiar." Our MIT pedigree garnered us grudging respect and got our feet in the door, but to the conservative world of investors, we screamed "high risk," "different," and even "dangerous."

After all, we were two "women scientists" (already outliers in the white, male world of investors), with one (me) in a hijab. We came from academia. Yes, we were both successful in our fields, but neither of us had launched a start-up or run a company before. And if that wasn't enough, there was one enormous hurdle to overcome: We were pitching an idea that made some men so uncomfortable, even hostile, that we had to describe it without actually saying what it was.

For several days, we had back-to-back meetings, morning until evening, with the top VC companies. Roz and I stayed at a hotel near the airport, and because I was still breastfeeding, Adam had come along for the ride. First thing in the morning, Roz and I would usually drop Adam off with a family friend from Egypt who took care of him during the day, but there were plenty of times that the sitter was unavailable, and we had to make do. Eight or so hours later, we would pick him up when we were done with our pitches, and then we often had dinner with prospective business partners.

The VC firms were all located on Sand Hill Road, a five-mile stretch in Menlo Park. We'd park our Toyota Camry rental next to the Benzes, Porsches, Range Rovers, and Maseratis driven by the VCs. (A few models were so high-end that I had never heard of them.) We'd then check in with the receptionist and be led into a conference room to wait for the VC investor to walk in. Often it was one partner accompanied by a younger associate. The VCs were always men; we never pitched to women or, for that matter, a person of color.

We avoided using the "e-word" (*emotion*) in describing our work, or any vocabulary that was too touchy-feely. In pitching our ideas, we used as many geeky tech terms as possible. The same men who recoiled at the word *emotion* were visibly enthusiastic about terms like *data points, mood-aware Internet, sentiment analysis, machine learning,* and *computer vision.*

We had our presentation down to a science: First, I clicked on the PowerPoint slide "Affectiva provides solutions focused on 'opt in' technologies that enable people to communicate their affective information and easily share this information with others and learn its meanings."

Click. Next screen: "Affectiva's core technologies include measurement of two key dimensions of affect, arousal (high/low) and valence (positive/negative) . . ."

We displayed our graphs and charts, and went over a laundry list of potential uses: autism, nonverbal learning disability, mental health disorders, sleep disorders, distance learning, new product feedback, call centers, online customer service, online social interactions . . .

A prototype is worth a thousand screen shots, so Roz was adamant that we pull out our demos early in the presentation. At MIT, she had studied people's physiological states—their level of arousal. Think of how alert and engaged you are when exposed to different stimuli. Roz found that people were the least revved up when passively listening to talking heads and most engaged during an interactive experience, when they could participate and ask questions.

Once we began the demos, even the biggest skeptics in the room looked at us with newfound respect. The guys would start making faces and watch the software track their expressions. Then Roz would slap the commercial version of iCalm, now called "Q Sensor," on our hosts' wrists, and they could see a graph of their fluctuating arousal levels. These demos always elicited great interest, laughter, and even fascination on the part of our hosts. But at least early on, no one took out their checkbook.

We must have pitched twenty VCs. Some were dismissive, but others, even if they didn't offer money, gave us some excellent advice. For starters, while we had considered our long list of possible uses for our technologies impressive, it was actually off-putting. The investors thought we were all over the place; over and over again, we heard from them the word *focus*. Given the fact that we were a company created around two very different products, software and hardware, focusing was easier said than done.

Indeed, we were being given a crash course in the facts of life of a start-up. As much as the VCs respected our good intentions to pursue technology for autism, they felt that they needed to see an application that would ultimately bring in a faster return on their investment. They were concerned that the "total addressable market" (TAM) for autism was too small in comparison to, say, that for a social media company like Facebook or Twitter, which appealed to everybody with Internet access. Autism, we were told, was just too "niche" a market. And there was the added hurdle of having to conduct clinical trials and get government approval before the product could be brought to market. We got the message: You can be a do-gooder up to a point, but the reality was that many investors wanted to make money and they wanted to make it fast. That meant that many of the uses we had in mind would have to be put on hold.

"THE TEAM"

When you're out pitching a start-up, one of the first things investors ask about is "the team." At the time, our team consisted of Roz, our two employees, and me. We didn't have a CEO—another strike against us. Given our lack of experience in running a company, Roz and I realized that we would need to hire a seasoned one. We interviewed quite a few candidates in Boston and were getting discouraged. One was a too-laid-back midlifer who seemed more interested in our vacation policy and whether we had a nightly cocktail hour for the staff than in growing the company. He didn't grasp our sense of purpose or our urgency to get moving, and we got the feeling he'd run the place like a country club. Another prospect didn't seem to get us or our technology at all but was still confident he could "bring us to new heights."

Finally, on one of our fundraising trips to Silicon Valley, an investor we were pitching suggested that we contact Dave Berman, who had been president of worldwide sales and services for WebEx Communications, which had recently been sold to Cisco. Dave was looking for the next big thing, and the investor thought this would be the perfect match.

We made a date with Berman to meet him for dinner. Once again, I couldn't find a babysitter, so Adam had to join us. I parked him in his car seat at our table. Dave, the father of three boys, was undeterred by the presence of a baby, and we hit it off right away. He was smart, driven, and ambitious, and really wanted to be our CEO. When we demo'd our technology, he was blown away. We talked about our commitment to ethical uses, and he nodded enthusiastically. The only caveat: He'd need to commute weekly to Boston from California, where his wife and sons lived. *Why not?* I thought. I was commuting from Cairo. Dave's commute would be a few thousand miles shorter. We'd make it work. So, we gave him the job.

To his credit, Dave turned us into a real start-up. He hired an executive team and someone to run our sales division, all based in California. The caliber of the people was amazing. We later hired Tim Peacock to run engineering. An MIT alum and computer pioneer who had led the development of Lotus 1-2-3, IBM's "killer app" in the 1980s, Tim would later become my chief operations officer and a trusted partner in running Affectiva. Unlike me, Roz did not resign her position at the MIT Media Lab when she became chief scientist at Affectiva, but she wasn't a day-to-day employee. I was our chief technology officer, overseeing the development of the core technology—and trying to spend as much time onsite as I could.

After all my schlepping from Cairo to Boston to Sand Hill Road and back, our first investment came from a group that knew and admired Roz from her work in the Media Lab. Peder Wallenberg, of the Wallenberg family, one of the wealthiest and most prominent families in Sweden, with vast investments in a wide swath of industries (from Big Pharma to electronics to engineering) and numerous philanthropic ventures, had visited the Media Lab a few years earlier. After seeing a demo of the projects being developed by her Affective Computing Group, Peder had mentioned to Roz, "If you ever need someone to fund your work, please let me know." After our Sand Hill Road adventure, Roz decided to reach out to him.

Intrigued, he responded immediately, and introduced us to Hans Lindroth, the managing director who runs Peder's fund, the Wallenberg Foundation. Hans is low-key and unassuming, but smart as a whip; you need to speak to him for only a few minutes before you realize how impressive he is. In any given week, Hans circles the globe, managing portfolio companies in China and nonprofits in Burundi and Brazil and conducting meetings with the queen of Sweden. He loved our technology, although he highlighted our need to focus; we clicked immediately. We had other suitors at the time, but what won me over about the Wallenberg Foundation was its strong presence in Egypt, where it runs several nonprofits, and its commitment to investing in Egyptian youth. I loved that—it meant it would support our decision to open an office in Cairo, something that was important to me.

Timing is everything for a start-up: If you're too early, you can fizzle out before your idea gains traction; if you're too late to market, you're always playing catch-up. Affectiva was in the right place at the right time. Yes, we were early when we spun out, but technology was moving in our direction. Fitbit was just coming to market; the "wearable space" was hot, but still in its early stage. The smartphone wave was under way, laptops were getting their first built-in cameras, and both these trends opened up the space for video-based communication. In turn, that meant that people were starting to be comfortable in the presence of computer cameras. So, asking people to turn on their webcams and use our facial analysis tracking system seemed a natural next step.

With funding now secured, we were poised to ride the wave of these new technologies.

16

My Arab Spring

I n 2011, Affectiva was requiring more and more of my attention; I was traveling to the States at least once a month, sometimes more. Most of the time, I still brought Adam with me, but Jana, who was now nearly eight years old, was in school and couldn't join us. I did try to speak with her every day, even if it was just for a quick "Hi, how was your day today?"

I spent countless hours on airplanes, on my laptop, keeping up with the business and technology news, but I was completely ignorant about what was happening in my own backyard. Little did I know that seething anger was brewing among the Egyptian population in the form of a grassroots movement against government abuse of power and, especially, police brutality. And this grassroots movement had a new tool for mobilizing: social media, in the form of Facebook and Twitter.

When I was growing up in the Middle East, most people were wary of politics: The process was so broken that my parents and I—really, anyone in my social sphere—never bothered to vote. Unlike

Election Day in the United States, Election Day in Egypt was not a big deal; I'm not even sure people knew when they were supposed to cast their ballots. Most of them felt it didn't make any difference. There was only one political party on the ballot, one candidate. The elections were riddled with fraud, and speaking out would only have invited trouble. The result? Generation upon generation of apathetic citizens who seldom got involved.

I was utterly clueless about what was happening in Cairo when, on January 20, 2011, I flew to Boston with Adam, who was twenty-three months old at the time. I was traveling with a young AUC graduate named May Bahgat, who had been a student in my Intro to Computer Science class at AUC a few years back. May was smart and ambitious, and I had recruited her for Affectiva's Cairo office. Now I wanted to introduce her to our Boston team.

We could use the extra set of hands. At Affectiva then we were "heads down," building the Q Sensor and ramping up our facial analysis platform in preparation for a joint Web project with Forbes Online that would run from March through April. We had invited the public to visit the Forbes site, turn on their webcams, and view some Super Bowl ads. While they watched, our algorithm would rate their smiles in real time. This project not only would enable us to collect more data, but could raise our profile among global marketers and brands. It was the first time that anyone would crowd-source emotion responses to advertising.

A few days into the trip, May, looking panic-stricken, pulled me aside and told me that something was going on in Cairo. Tens of thousands of people were protesting in Tahrir Square, near the AUC campus. I hadn't heard anything about it from Wael or the rest of my family. Then May showed me her Twitter account. There were thousands of live tweets from the *midan* ("square" in Arabic). I naïvely shrugged it off. Mubarak's government would break up the protests, I said. And I went about my work.

The next day, May seemed even more worried. Her parents had urged her to fly home. This meant I had to fly back with her; even though she was in her mid-twenties, smart, and independent, she was not allowed to travel by herself. I felt that her parents' concerns were unfounded, and I tried to persuade her to stay, thinking that what was happening back home would fizzle out. But it didn't. The protests continued to grow in size—the news commentators dubbed the movement "the Arab Spring."

The following day, all flights to Cairo were canceled. The airport in Cairo was shut down. Schools were closed, offices shuttered, and a three P.M. curfew was imposed on the country. My mother called me; my youngest sister, Rula, was among the protesters, and my mother was sick with worry. At first Rula had kept her actions secret from my parents, turning her phone off when she was in Tahrir Square so that she couldn't be located. But with the demonstrations stretching on for days, it was impossible to hide the fact that she was there protesting. My mother feared for my sister's life, as did my father (though, he was also secretly proud that she had the courage to stand up to Mubarak's corrupt regime).

But the bad news kept coming. Convicts, violent criminals, had escaped from the country's prisons and were attacking people in the neighborhoods of Cairo. My uncles and brother-in-law, armed with knives—in Egypt, no one other than the police and the military has access to guns—were taking turns guarding our houses.

I frantically called Wael. "Bring me Jana. Get on a plane out as fast as you can." I wanted my daughter with me; I didn't know how long the protests would last. He assured me that she was fine; he had sent her to his parents, who lived outside the city, in the suburbs of Cairo. My father-in-law was homeschooling her for the time being, while my mother-in-law took care of her. Together they cooked food that my mother-in-law sent to support the protesters in Tahrir Square.

At the time, Jana was learning how to use Scratch, a visual programming language developed by Mitch Resnick and his team at the MIT Media Lab's Lifelong Kindergarten Group. I was a huge fan of Mitch's work, and had gotten Jana interested. My in-laws had a large-screen TV with the news on 24/7. While she was out of school, Jana built a program that retold the news on Scratch; I think it was her own way of attempting to process what was happening.

By Saturday, the phone and Internet service to Cairo had been cut off. I had no way of finding out how Jana or the rest of my family was. Fear is a terrible emotion, especially when you feel helpless to do anything about it. And I knew there was nothing I could do from Waltham, Massachusetts.

For a distraction, I did what I always do: I threw myself into work, where I had control. I couldn't control Egyptian politics, but I did hold some sway over Affectiva's future. I emailed the team (fifteen of us at the time) and announced that I wanted everyone to meet at the office to plan our 2011 strategy. It was a Saturday morning, the weekend, yet everyone showed up. We met in our large conference room. I gave Adam a few toys to play with at the far end of the room.

Where did we stand on our upcoming shipment deadline? I asked. Where were we with testing the software for the Forbes project? I knew we needed to raise more money at some point in 2011. Which investors should we reach out to?

By Sunday, I still had had no communication from home. I began to imagine the worst: What if violent thugs had stormed our house? What if members of my family had been hurt?

On Monday, telephone service between the United States and Egypt had been restored. I sobbed with relief when I managed to speak with my family and was told that despite the turmoil, Jana was not only doing well, but my in-laws were keeping her safe and happy. To this day, those two weeks of homeschooling are Jana's fondest memories of her grandfather.

When Cairo International Airport reopened, I booked seats on one of the few flights to Cairo, and on February 2, May, Adam, and I flew from Boston to Frankfurt. The flight was full—every seat was taken—but when we transferred in Frankfurt to the flight to Cairo, the plane was eerily empty. There were only two other passengers in the cabin.

When we landed and exited the plane, I saw why: The gate was swarming with people, Egyptians and tourists alike, desperate to get out of the country.

I remember the long taxi ride to my in-laws' home. We were nearing the three P.M. curfew the government had set, and the driver was racing to get us there in time. The streets were empty except for a massive number of soldiers and tanks. But when I got to my in-laws' and held Jana safely in my arms, nothing else mattered.

Over the next few days, I watched the revolution on the streets of Cairo play out on our television set. Hundreds of thousands of protestors continued their siege of Tahrir Square, including my sister. I was so preoccupied with what was going on that I couldn't work. Finally, on February 11, Mubarak's regime was toppled and power was transferred to the Supreme Council of the Armed Forces, the Egyptian military. People in Egypt were jubilant. The next day, we went back to our home and painted the curbsides with the colors of the Egyptian flag: red, white, and black. Everyone seemed to be out on the streets waving flags. There was a palpable sense of hope and community.

A few weeks later, with the crisis ended, I returned to Boston to find myself in the throes of another kind of crisis, but this time one with dire consequences for my company. Our seed investment of two million dollars was running out. We were burning through our investment money at a rapid rate. We needed to raise an additional five to seven million to be able to grow the team and invest in building the technology. There was already a great deal of interest among companies in using our technology for market research; this seemed to be the logical place to turn for funding.

For decades, market research has in large part relied on focus groups, panels of prospective consumers who typically meet in person and, under the direction of a moderator, are asked to rate a product (like a beverage or a TV show), commercial, or even a political candidate's debate performance. Focus groups, however, are fraught with problems because human beings are, well, complicated and, at times, biased.

Self-reported data is itself flawed; we all want to be liked, and participants sometimes offer what they think the moderator *wants* them to say, as opposed to their true, gut response. Sometimes the sample itself is skewed. Think about it: How many people today actually have time to participate in a group? If you're employed, or raising children, or if you're a full-time student, or just busy with life, you may not be willing to sacrifice an afternoon or evening to participate in a focus group. That fact alone limits the diversity of the group.

This is not to say that the data collected from conventional focus groups is necessarily wrong or skewed; it can provide valuable insights. But it doesn't always tell the whole story. That is why products or commercials that test well among focus groups, ones that, based on the data, should be surefire winners, may flop in the marketplace. And after all, if you're spending millions to produce an ad to run during the Super Bowl, you want to make sure that it sells the product and doesn't offend any of your potential customers.

We offered a new and novel alternative to the conventional focus group. As smartphones equipped with state-of-the-art cameras became ubiquitous, it became possible to collect data from a diverse population anywhere in the world—where people worked, lived, and played. It was as simple as sending a link to the smartphone or computer of potential participants and asking them if, in exchange for a modest payment, they would watch an ad and allow us to record their reactions with their own cameras or webcams so that our facial decoder could analyze those responses. It is an easy way for someone to pick up an extra five or ten dollars, and a scalable way for us to gather massive amounts of data. And it is always done with the consent of the participant, and sometimes we even offered to show them their personal results.

Just from asking people to participate in the Forbes project, we gathered 3,268 videos. At the time, we had built the biggest database of naturalistic facial responses ever collected. Remember, these weren't actors; these were ordinary people who came to our website voluntarily, perhaps out of curiosity. But it was also a remarkably diverse group.

The Forbes project caught the eye of Millward Brown, or MB, now Kantar Millward Brown, a division of WPP, a major international branding and marketing company. MB was willing to invest seven million dollars for access to our technology, money we desperately needed. But first they wanted us to use the algorithm on four commercials they had already audience-tested using conventional means: Dove's "Onslaught" ad; Huggies's "Geyser"; a spot for Lynx antiperspirant, by Axe; and an ad for BMW. MB already knew whether the ads had worked or flopped, but they wanted to see how well our algorithm performed, whether it would accurately identify when viewers smiled, frowned, looked engaged, and so on—and whether it could offer even more insights.

We felt comfortable with the people from MB. Unlike some of the companies that had approached us, they agreed that they would not use our technology without the consent of the participants.

Each member of our team reached out to up to five of our family and friends to watch the ads. Fortunately, we are a diverse group, so our sample was equally diverse. This first round of testing was designed to make sure the experience of viewing the ads worked on multiple laptops and browsers. We had not yet designed our technology to work on mobile phones—this was still early in the mobile revolution. When we were confident that everything worked well, we recruited volunteers via the Internet, offering five to ten dollars to anyone willing to watch the ads and allow us to observe their reactions via their webcams.

The Huggies ad was the most successful in terms of connecting with consumers and maintaining brand identity. But, to me, the viewer results from Dove's "Onslaught" ad were by far the most interesting, perhaps because I am the mother of a daughter, and the ad focuses on girls and self-esteem.

The Dove ad kicks off with a shot of a strawberry-blond girl of about eight or nine. The viewer is then quickly bombarded with images of skinny models with impossibly fit bodies gyrating in bikinis, advertisements for products that promise to make you "younger, smaller, lighter, firmer, tighter, thinner, softer." It culminates in a series of images of women prepping for cosmetic surgery and breast enhancement. The ad was hard to watch, and elicited very strong negative reactions, especially from women.

The most unforgettable part of our outreach was a video of a female viewer displaying a grimace of disgust as she watched the Dove ad, with every lip curl, brow furrow, and nose twitch quantified in vivid detail. It was a perfect depiction of what Emotion AI can capture.

The Dove ad concludes with a scene of young preadolescent girls walking confidently across a street, a lead-in to the next screen shot: "Talk to your daughter before the beauty industry does." It is followed by a screenshot that tells the viewer to download the company's self-esteem programs from a website; a final screenshot identifies the sponsor, the Dove Self-Esteem Fund. The problem was that, by the time of the final shot, many people had assumed the ad had ended. The brand reveal came too late.

Although the ad was hugely successful in eliciting negative responses from viewers, as designed, and although it was critically acclaimed—an *Ad Age* reviewer said it should win an Oscar for "really, really short subject"—it didn't really work as intended. The cosmetic surgery scene elicited extremely strong expressions of disgust and aversion, but there was not enough time for the viewer to bounce back after those disturbing images. Also, the ad lacked a feel-good ending, and perhaps because it didn't end on an inspiring note, as powerful as it was, it didn't go viral the way other Dove ads had.

With each of the ads, we delivered to MB insights it could never have gotten from a focus group or a survey. We were able to map out in real time the viewer's emotional journey, second by second, pinpointing the subtle shifts in the viewer's mood. We could track both the intensity of emotion and viewer engagement levels.

By April, as we were furiously gathering data for MB, we had only two months' worth of cash left in the bank. Soon, we wouldn't be able to make payroll. We were still in discussion with WPP, MB's parent company, but as with all fundraising, getting funded takes a lot longer than you think. One gloomy day, we got a cold call from the ventures arm of an intelligence agency. They were *very* interested in funding our company. They believed there was a huge opportunity for our technology in surveillance and deception detection and wanted to fund us to explore that space. They offered us *forty million dollars*, a staggering sum for a young start-up like ours. It would allow us to hire and grow and would give us several years of "runway" before we would need to raise money again.

So, we had two clear choices, and they couldn't have been more different. Option A: Take the funding from the government and shift our focus to security and surveillance—and probably make a ton of money in the process. Option B: Walk away from the money knowing that we might run out of cash by July and have to shut down the company.

I went home and tossed and turned all night, thinking about our options. I desperately wanted our company to grow, to survive, but I couldn't imagine spending my hours working on software that focused on spying on people. I wanted us to be the company that people entrusted with their information. But how could we be that trusted partner if we sold their data to the government? I had just witnessed in Egypt what can happen when a government becomes oppressive and tramples on the rights of individuals. I didn't want our technology ever used in that way.

I recalled the conversation Roz and I had had in her home in early 2009, when we laid out the core values that defined who we were as scientists and human beings. And when I did, the decision was easy. Affectiva was about trust and respect for people's privacy; those goals were just as important two years later. I walked into our CEO Dave's office and told him that we could not take the government money. We would have to double down on other potential investors waiting in the wings.

For the next two months, I went to work every day with the burden of not knowing whether we would survive. I'd look our employees in the eye and wonder, *Will we be able to pay them?*

The negotiations went down to the wire. In May, as we finished our work for MB, Graham Page, the executive sponsor of the project, a Brit and an Oxford graduate, flew to Boston to meet with us.

We shared our results, highlighting what had gone well, but also where the technology had fallen short. For instance, there were cases where people grimaced and we incorrectly classified that as a smile. We also showed him the video of "the face," the grimace/disgust reaction to the cosmetic surgery scene that the participants had allowed us to share with MB. It was a very powerful reaction and statement. I think that video and the data, along with our team's transparency, were what tipped the balance in our favor.

Seven days before we were due to run out of cash, MB signed on, and we brought in a seven-million-dollar investment. Not as much as forty million, but enough to give us the extra runway we needed. More important, we would be working with partners we loved and who were aligned with our mission and our core values.

With money in the bank, we now grew the team, hiring more machine learning scientists and software engineers, who ensured that our facial platform ran flawlessly 24/7. Now that we were out of academia, I could no longer hold my algorithm's hand, so to speak, coaxing it every time it faltered. And for the most part, it was performing well. On one very important occasion, however, our technology stumbled.

GOING GLOBAL

MB began integrating our software all over the world, one country at a time. Then it hit a snag. When you're trying to keep a start-up afloat, even a minor problem can seem like a major one, and a major problem can feel like an existential threat. One day, we got a panicky call from the senior partner at Millward Brown informing us that our technology "doesn't work in China, and China is our biggest client!" At the time, MB was testing advertising for Fortune 500 companies doing business in China.

I was stunned. What was going on? I was also deeply worried. If we were to succeed as a global company, our algorithm had to be able to perform in China, a country that represented one quarter of the world's population. Failure to solve this would be the end of Affectiva's hockey stick–curved growth.

I asked the team to collect all the data on China that MB had amassed so far, so I could study it. I watched each video, frame by frame, trying to understand why the algorithm didn't seem to "get" Chinese viewers. After hundreds of videos, I started to notice a pattern. When Chinese participants were watching ads with the researcher standing next to them, their emotions and facial expressions were very subdued—in fact, practically nonexistent. But when they were watching the ads alone, they were highly expressive, on par with what we saw in the States.

I dove into the research literature. From my PhD work, I recalled the cross-cultural differences in how people express emotions. In collectivist cultures like China, cultural norms amplify or mask one's true emotions. And these norms kick into action most strongly in the presence of strangers.

As I watched the video, I noticed something else that could be skewing our results. Many of the Chinese test subjects wore a smile as their baseline expression; an ever-so-slight lip corner pull. A naïve observer might think this was a smile expressing happiness, but I knew better. It was the smile of politeness that I had often used myself as the "nice Egyptian girl," the smile of a man or woman who didn't want to offend anyone, the play-it-safe smile.

Having analyzed millions of data points from our research, we know today that this smile of politeness, this social smile, is far more prevalent in collectivist cultures than in individualistic cultures like that in the United States. We therefore needed to change the protocol in China so that participants watched the ads alone, out of the view of the researcher, so that they felt comfortable expressing their true emotions. We also went back and added dozens of additional examples of that politeness smile to our training set, so that the algorithm could distinguish it from a smile of genuine happiness.

A week later, the executive from MB called me back. He was jubilant: Our China problem was fixed, and our customers were happy. Everyone exhaled. Crisis solved.

The takeaway here is that despite how sophisticated AI is as a tool, it was, after all, human beings (my team and I) who, using life experience and intuition, were able to solve the problem. And those are skills that will never become obsolete.

Through our work with MB, our tech was now deployed in ninety countries, and our database had grown to include millions of facial responses. It vastly expanded both the insight (EQ) of my algorithm and, very quickly, our understanding of how emotion is displayed among people of all genders, ages, nationalities, and ethnicities. If I had remained a research scientist at MIT, I may not have had the funding or the staff to do the work at this scale.

17

Grounded in Cairo

With our new partnership with WPP, Affectiva in 2012 was moving forward, but work was all-consuming, requiring my hands-on attention practically 24/7 ... even on vacation. That summer, my family and I decided to unwind with a one-week trip to Los Cabos, Mexico, where we stayed in a lovely condo facing the Pacific Ocean. Nonetheless, I found it hard to leave my work behind. I monitored email constantly and jumped on conference calls throughout the day.

While on vacation, it suddenly dawned on me that Wael and I hadn't had sex since the time we'd slept together to conceive Adam, which was more than three years ago. In reality, we really hadn't been intimate for two years before that. How was that possible? How could someone who was an "emotion expert" have missed such a huge red flag? Wael and I had gotten in the habit of sleeping in separate beds; that had started when the kids were young and wouldn't sleep alone, and it sort of stuck. I'm not an expert on marriage, but perhaps this happens more often than I realized.

One evening, the kids went to bed early, and Wael was on the couch watching TV. The sun was setting; it was a beautiful evening. I felt relaxed and romantic. Sitting next to him on the couch, I said playfully, "You know we haven't had sex in a while. What do you think?"

His response was like a punch in the gut. "Rana, you're kidding, right? Where have you been? I'm done with our marriage. I want out."

I saw anger in his face, even contempt. Wael said that he was not a top priority in my life. And that was how I learned that Wael wanted a divorce. I was, to put it mildly, devastated. I was clueless that he felt so alienated, so unhappy. While I was crisscrossing the globe, sharing the secrets of teaching computers how to decode human emotion, I had missed my own husband's emotional cues. It was ironic. I had not paid attention to how little time Wael and I spent together as a couple. And it wasn't just because of my absences. Even when I was home, I realized, I was often distracted with running Affectiva.

Our extended periods of living apart, exacerbated by the demands of a growing start-up, had taken a fierce toll. Since the launch of Affectiva, my schedule had been beyond insane. Between the company and my children, I hadn't had much time to breathe. Wael had a point. In my life, work and kids had come first. I took it for granted that he and I had a close relationship and that our marriage would endure. I operated on the assumption that if Wael focused on his work, and I focused on my work, and we focused on the kids together, all would be fine. Clearly, it wasn't fine with him. He missed feeling cherished, supported, and loved.

I was also moving in a new, uncomfortable direction for Wael: I was no longer an academic. Like him, I had my own company to run. And it was working—Affectiva was getting noticed. VCs had invested millions of dollars in the company I had created. Now my dreams were focused on developing its full potential. Had I stepped on Wael's toes in the process? Had I encroached on his entrepreneurial territory? Was I no longer a partner in our marriage, but rather, a competitor?

I think Wael figured out long before I did how much I had already changed. I was no longer the innocent young Egyptian girl he had married. And my vision for the future was sure to ultimately collide with "our vision."

Perhaps Wael put it the most succinctly: "You don't need me."

For some time, we had pretty much been living separate existences, but when we were together, neither one of us wanted to acknowledge this. We never argued, and we were always respectful of each other. But our marriage had become very transactional, dealing primarily with the business of family. The romantic spark had burned out. That realization shook me up, badly. Wael, the father of my children, had always been my best friend. I believed that we could reconcile our differences, if we worked at it. I suggested that we meet with a marriage counselor, but Wael wouldn't hear of it. Wael was clear: He'd had enough. He wanted out. It was over between us. Divorce was now a formality.

The word for *divorce* in Arabic is *talaq*. It was a word that neither Wael nor I had ever wanted to utter in front of our parents. We now dreaded the prospect of telling them, so we kept quiet about our situation for a while. In public, we looked like the perfect happy, modern, two-career couple with two great kids, living the good life. No one knew how broken our marriage really was. So, it came as a complete shock to our parents when we finally summoned the courage to tell them that we were getting a divorce. Both sets of parents were angry, even furious. Our mothers were heartbroken—they couldn't even talk to us without crying—but Wael's father and my father met with us and laid down the law. They told us that divorce was completely out of the question. We would disgrace our families and ourselves. Our professional lives would suffer, our children would be emotionally damaged, and just about every calamity known to man would befall us. They pleaded with us—more like ordered us—to take whatever steps were necessary to patch up our marriage.

Wael's older brother, who lived in the United States, flew back to Cairo to help us find a way to reconcile. I definitely was on board; I wanted to make us work again as a couple. I missed the closeness that we once had. Wael agreed to try, but it was clear that he was really being badgered into something that he didn't want. Obedience to parents had been drummed into our heads from childhood. Even as successful adults in our thirties, we succumbed to the pressure. And part of me didn't want to give up on our marriage because I didn't want to be viewed as a failure. Out of love, duty, respect, and fear of the consequences of divorce (and what people would think of us), we crumbled. Wael and I decided to accede to our parents' wishes.

Our parents placed much of the blame for the unraveling of our marriage on me. I was the nontraditional woman. I was the entrepreneur who commuted to America. To both sets of parents, the solution to our marital woes was simple. I needed to be a better wife. And I went along with it. The older and wiser Rana of today understands that it takes two to tango. I wasn't the only one who didn't nurture this marriage. Wael never expressed any misgivings or talked openly about his feelings to me. But nice Egyptian girl Rana succumbed to the voice in her head that kept saying over and over again, "It's all your fault. You did this to yourself. You're such a failure." And so, I listened and obeyed when Uncle Ahmed and my father laid out the ground rules to save our marriage. Uncle Ahmed urged me to become a better cook because "the fastest way to a man's heart is through his stomach." My father insisted that I stop traveling to the States, give up running Affectiva, and focus on my home and family. The cooking part I could deal with, but I resisted the travel ban. I was being asked to turn my life upside down, and Wael didn't have to change at all. My father and I argued over my future back and forth for nearly a year. By early 2012, I was so worn down that I succumbed to family pressure and agreed to stop traveling to Boston and stay home in Egypt full-time to try to save my marriage. I jokingly told the Affectiva team and my board that I had been "grounded" in Cairo and would henceforth have to work remotely.

I was crushed by Wael's rejection, and I dealt with my feelings of hurt and loss the best way I knew how. I attacked it with a problem-solving mindset: I devised an action plan.

Top of the list was Project Rana. Wael's desire to abandon our marriage was an enormous blow to my self-esteem. I became fixated on my physical appearance. Only thirty-four, I thought I looked washed-out and drab, overweight from the baby pounds I was carrying from my last pregnancy. I felt as if I was no fun anymore, unsexy, even old. Juggling motherhood and the start-up had made it impossible to keep up with the latest fashions, or to worry about accessories, or to get myself to a gym. Now that I was working from home in Cairo, I had more time. I joined a gym, toned up, and slimmed down. I retired my dowdy wardrobe for trendier clothing styles. It was as if I were trying to re-create my identity from the outside in.

I was adamant about maintaining my physical and mental health. I needed to lift my spirits, so I got into the habit of watching romantic comedies on my iPad while working out on the elliptical machine. It helped jump-start some laughs. My all-time favorite rom-com is *The Holiday*, a film about two women, one from a quaint British town and the other from Los Angeles, both unlucky in love. They swap homes during the Christmas holiday and meet the loves of their lives. I watched it over and over again; clearly, I was looking for a happy ending. I also made it a point to be around people. I got very involved in Jana and Adam's school and volunteered to be on the school board.

I made a tremendous effort to rebuild our family life. I honed my cooking skills, which weren't much to begin with. One evening I summoned the family together to make sushi, which Wael wryly referred to as "team building." And when Wael was away on a four-day trip to Dubai for business, I decided to redecorate the bedroom in an effort to dispel the bad memories so that we could start fresh. With the help of my mother and my aunt, I selected new wallpaper and a lovely bedding ensemble, and decluttered the room. When Wael came home, I showed it to him with a big smile. He nodded in approval. Up until that point, I had been sleeping in Adam's room. I asked, "So may I sleep in here?" "Absolutely not!" Wael responded. I hid my disappointment and swallowed my tears.

To be fair, Wael didn't lead me on; he told me that he was sticking around because his parents had asked him to stay, and for the kids. In other words, he was telling me to stop trying to fix something that couldn't be fixed.

After being grounded in Cairo for several months, I asked for a brief time-out to fly to Boston to meet with several key clients. I undertook the trip only because it was important for the future of the company. As I was sitting at my desk in Boston, poring over data, my father called. "Rana, forget about Affectiva! Just sell it! Or resign. Get rid of it! Tell them that you can't work there anymore."

"Dad, what are you talking about, *it's my company!*"

I knew he was coming from a place of love—his priority was fixing the marriage because he believed that would ultimately make me happy—but I was still disappointed. It was also clear to me that the same rules didn't apply to men. No one was asking Wael to leave his company to move to Boston. Why not? I was also deeply hurt that my dad was asking me to walk away from what I'd worked my entire professional life to achieve, and just when our fledgling company seemed to be hitting its stride.

It was hard to accept. At the time, I believed that my parents didn't understand what I did. I felt that they didn't feel proud of me. If they could choose, I felt, they'd pick Rana the happily married mother and housewife over Rana the successful AI entrepreneur. What *I* wanted wasn't important to them.

I never shared with anyone what my father said to me on that phone call. Nor did I talk to my dad about it again. But his point had been made, and I've never forgotten it.

Even as I was hiding out in Cairo, my work was getting noticed. In September 2012, I was selected by *MIT Technology Review* as one of its "35 Innovators Under 35" in technology. It's a very prestigious list to be on—among the people who have made it are Facebook founder Mark Zuckerberg and the cofounders of Google, Sergey Brin and Larry Page.

I couldn't go to Boston to accept the award because I was grounded, so my mom threw a small surprise celebration at her place, inviting my sisters, my aunt, and my in-laws. She prepared a small feast and bought a cake. When I got to my parents' apartment and realized what she had in mind, I was alarmed. Such a party would only exacerbate the tensions between Wael and me.

"Mom, please don't do this," I said. "Don't mention my award. Let's just make this a family dinner, okay?"

My mother got it. I didn't want to draw any more attention to the fact that I was becoming more and more successful. It would further reinforce Wael's belief that I had prioritized work and my career over him. I could feel the fracture lines it would create in my marriage, and I was determined to underplay my success and achievements rather than hurt Wael. Nevertheless, I was bitterly disappointed not to be able to accept my award in person.

RANA 2.0

In the midst of all this angst over my marriage, I made an impetuous decision. On a brisk, sunny December morning (Christmas Day, in fact), I prepared to do some grocery shopping. I could hear Jana and Adam playing downstairs as I put on my jeans and a sweater. I automatically pulled out a headscarf from my dresser from among the many I owned. And then, just as I was about to tie it on, I didn't. I left it on my dressing table, went downstairs, grabbed my purse and car keys, and headed out the door. Adam, who was three at the time, noticed immediately. "Mummy, you forgot your headscarf," he said. I paused and told him that I didn't need it. He had never seen me go outside without a scarf in his life. He looked upset and a little confused. At that point, I dashed out, afraid I'd lose my courage.

I rolled down the windows of the car and sped onto the highway. I wanted to feel the sensation of the wind in my hair, something I hadn't experienced in the twelve years since I first put on the hijab. I turned on the radio and cranked up the volume.

On the surface, the removal of hijab may look like an act of "rebellion" against a culture that treated Wael and me differently, that seemed so determined to hold me back. In part it was a reaction against the unfairness of the situation. But it was also more complex. Some of it had to do with pure vanity. Honestly, the hijab was making me feel old. I wanted to turn back the clock and be that young girl again with flowing hair, who was fun and "with it," not an older woman whose best days were behind her. I not only felt physically unattractive, but I came to believe that my personality was as dull as my exterior. At the time, I hated myself. Removing the hijab was a way of loudly announcing, "Hey, this is Rana 2.0, the fun, cool Rana!"

But without question, the political climate of the time also influenced my decision. Removing the hijab was a minor act of resistance against the Muslim Brotherhood, the reactionary religious party that had won a nationwide election by promising reform. Egyptians were eager for an end to the chaos that ensued after the end of Mubarak's regime. But the Muslim Brotherhood seemed hell-bent on turning back the clock on women's rights, disappointing many of the young people who participated in the Arab Spring demonstrations. The sudden loss of power of women in Egyptian society was not going over well with the women I knew, even the religious women. The mantra among family and friends was "We don't want to become another Iran!"

But on a deeper level, my views on religion—my worldview—had changed. I no longer felt that outward displays of religious devotion were the sole measure of one's spirituality. The true test of faith was how you conducted yourself in the world, and the respect and kindness you showed others, the empathy you displayed. "Wearing your religion" no longer seemed to matter to me.

That didn't mean that I had become any less of a Muslim, or that I don't respect the hijab—my mother still wears one, as does my middle sister. And many of the women who work for me wear hijabs. They are smart, informed individuals with their own points of view on how they practice their faith. But for me, the hijab no longer represented the person I had become.

A few days later, I launched the new, improved version of Rana. During the holiday week, I threw a gingerbread house making party for Jana's friends and their mothers. This may sound like an odd thing for a Muslim girl to do, but even in some Muslim countries, Christmas and New Year's are observed as secular holidays. I wore very fashionable clothes, showing off my new svelte figure. You know what? The new Rana felt lighter and funnier than the old Rana, and I had a blast. For a few hours, I forgot that I was stuck in a miserable marriage with a man who didn't want to be with me.

Yet, with or without my hijab, I was still, at heart, the nice, obedient Egyptian girl I had been raised to be. I worked around my family life, just like my mother did. My week was divided between Cairo days and Boston days—I alternated between the two. Every morning I would get Jana and Adam ready for school; on Cairo days, I would then head to Affectiva's Cairo office to work with the team there. At three P.M., I would race home to see the kids when they got off the school bus.

On Boston days, I worked out of my home office, which was basically a big armchair facing the main entrance of our house. Wael would sometimes return home at night when I was on a conference call or having a heated discussion with the team. My heart would sink as he walked in. Even though he knew that I tried to limit my contact with the Boston team to every other day, I felt guilty working when he was at home. I did not talk about Affectiva or business when the children and Wael were there, and I dutifully complied with a code of behavior set by my father (just as my mother had done).

Still, I couldn't separate from Affectiva completely, or ignore when we had a problem, one that could ultimately put us out of business if we didn't solve it. All start-ups have growing pains, and we had our share. One of them was establishing our identity, our brand. Founders need to have a simple "elevator pitch" to describe to investors and others what their company does. The problem for Affectiva was that we seemed to be stuck between floors.

We were trying to promote two entirely different products: Roz's Q Sensor, which pushed us into the hardware business; and our software algorithm, which at the time we called Affdex. It is difficult trying to get any start-up off the ground, but trying to tackle two markets simultaneously caused a mini rebellion among our sales staff. The running joke at the office was that we sold to two types of "CMOs": the Q Sensor to chief *medical* officers for clinical research, and the facial analysis platform Affdex to chief *marketing* officers for advertising research.

Ultimately, we had to make a choice. What kind of company were we? Software is a much easier and more profitable product to sell. It exists in the cloud and runs on computers that people already own. In contrast, we had to manufacture the actual Q Sensors as well as sell and ship them. As with any kind of hardware, a lot can go wrong. To compound the problem, our core competency as a team was not in manufacturing. And because the wearable sensor was focused primarily on the health market (for pain, epilepsy, and autism), it required clinical trials before it could be approved as a bona fide medical device.

Bottom line: We were achieving a profit margin of about 90 percent for our software and we simply could not make the business case for the hardware.

Nonetheless, the Q Sensor was Roz's passion. It was her reason for having spun out of MIT. She hoped to use the technology to create the next generation of tools for health. I agreed with her vision, but my immediate focus was ensuring that we successfully transitioned from a lab experiment to a viable, sustainable business that drove revenue and put us on a path to profitability.

On a chilly spring day, the Affectiva board voted unanimously to discontinue the hardware business. It was tough, but it was the right decision. The data was clear. To continue to exist, we had to sunset the hardware. It was the first of many hard decisions I would have to make on the path to becoming a viable business.

If I was upset about killing off the Q Sensor at Affectiva, Roz was even more so. Just as Affdex was my baby, the Q Sensor was her brainchild, and in true Roz style, she remained determined to bring it to market. In March of the next year, she formed another company to market the Q Sensor and develop "consumer-friendly wearables with clinical quality data." A year later, her company merged with Empatica Srl, and the parent company was renamed Empatica Inc.

Amid all this turmoil in my life, there was yet more unsettling news. In May 2013, I was sightseeing in Istanbul with my children and extended family when I suddenly received an email from Dave, our CEO at the time, asking me to call him right away. With my heart in my throat, I called him from the hotel line. I was relieved to hear that everything was fine at Affectiva. He was calling to tell me he had accepted a job offer with another company.

It was a possibility I hadn't seen coming. At first, I was in a panic. What did this mean for Affectiva? Who would run the company?

REGRETS

A few weeks later, on a conference call with the Affectiva board, I was told that, much to my astonishment, the board was considering me as a potential replacement for Dave, whose departure was imminent. The two women on the board at the time felt I should be CEO. "Rana knows the technology inside out. It's her baby."

That evening, I mentioned to Wael that the board was considering making me CEO. Without missing a beat, Wael tossed off, matter-of-factly, "Oh, it would be a great career move for you. Detrimental for the company!" He didn't say, "Oh, what are your other options? Let's talk this over," the way he used to. He just walked out of the room. His casual comment hurt me more than I let on. It was yet another blow to my self-esteem.

So I hesitated. I had never been CEO before, and as I've learned through the years, women tend to only raise their hands if they check 110 percent of the requirements. Men, well not so much. On the call, Nick Langeveld, our vice president of business development, leapt at the chance to be CEO, even though he, too, had never been a CEO. Nick became interim CEO, and within a couple of months, we voted him in as CEO. I didn't fault Nick for stepping forward, but I was angry with myself for not having the confidence to take the risk.

Later that night, I burst into tears over what I had done. I wrote in my journal, "This doesn't feel right. This was a mistake. I know I'll regret it."

A few weeks later, I turned to Wael and said, "We aren't making any progress, are we?" He shook his head. "No, we are not. It's time to end this." After a year of my being "grounded in Cairo," it became apparent that Wael wasn't interested in salvaging our marriage. And I was done pretending to be someone I wasn't.

Even if I'd wanted to, I couldn't roll back the clock and play the part of submissive wife and mother with a career on the side. It was unsustainable: That was not who I was. Neither Wael nor I was happy, and things were not getting better. I missed Affectiva, and I missed Boston. I certainly didn't want to lose my company and my marriage; at least I had some control over my company. So, we decided to separate and eventually go ahead with the divorce, over our families' objections.

I had to decide what my next move would be: Would I stay in Cairo with my family and continue to commute to Boston, or make the break and move to the States with my kids to make a new life for myself in Boston? Could I do it on my own?

That summer, I took Jana to Belgrade, Serbia, to compete in the Petar Konjović International Competition that attracts young harpists from all over the world. I know Jana enjoyed the trip and got a lot out of the weekend, and she won second prize in the competition. But going to Serbia had an unexpected influence on me. I found myself deeply moved by the words of Irina Zingg, chair of the program, when she told the group of young musicians, "When you push boundaries, you grow."

The phrase hit home: It was *my time* to take a risk, to push boundaries, to grow. Later that day, I told Jana that things between her father and me were not going well, and I was contemplating moving to Boston. Of course, I would take her and Adam with me. At first, she cried, but true to form, she pulled herself together and then went online to figure out where we would live and where she would go to school.

Just as my life was in turmoil, so was Egypt. Things had gone from bad to worse, quickly. President Morsi's unpopularity was growing, and the military intervened, threatening to oust him on the grounds that he had been accused of usurping too much power. Egyptians, especially women, were fed up with the Morsi regime. Tens of thousands of women, some in jeans, some in hijabs, including my mother, were taking to the streets to protest and support a military coup against the Brotherhood. (Egyptian politics can be *very* complicated.) Once again, people were looking for a return to "normalcy."

By the end of June, the Morsi regime had been toppled, and Egypt was in chaos. I made the decision to leave, but there was still one hurdle to overcome: Under Egyptian law, I would need Wael's consent to take the children out of the country. He agreed with my decision, recognizing that there were greater educational and economic opportunities for Jana and Adam in the States. Our parents knew the truth, but Wael and I kept our plan to get a divorce from our friends and even from Jana and Adam. Since Wael and I had lived apart so much of the time, this arrangement didn't seem odd to the children. And since they had spent so much time in the United States, it felt like their second home. Wael promised that he would visit them in Boston (and he did), and I would bring them back to Cairo for vacations. Two weeks later, the five of us—Jana, Adam, my mom (who came along to help us get settled), our white Persian cat, Claudie, and I—flew to Boston.

The day before we left, we said our goodbyes to my in-laws. Despite the problems in the marriage, we had remained very close. My father-in-law treated me like a daughter; he begged me and the children not to go. He feared that our lives would be forever ruined. He could be tough and demanding at times, but he was also very supportive of me. I think he understood my ambition, maybe even better than Wael did. When he hugged me goodbye, I saw tears in his eyes. I don't think that I had ever seen him cry before. He suffered from a heart condition, and I knew he was worried that he would never see us again.

Two days later, back in Boston, Jana and I were walking through a shopping mall when I got a text from Wael's sister-in-law to call her. We sat down on a bench, and when I got her on the phone, she delivered some terrible news: Uncle Ahmed had died of heart failure. I was stunned. I told Jana, and she was devastated. I felt a mix of sorrow and guilt—Uncle Ahmed once told me that he would rather die than see Wael and me apart. I couldn't have felt any worse.

I immediately flew back to Cairo to attend his funeral. My mother-in-law was very warm to me, as was the rest of Wael's family. They knew how close I had been to Uncle Ahmed.

It was a dark period in my life. I felt alone, completely abandoned. In less than a year, I had lost my support network: Wael and I were now officially separated, headed for divorce; Roz had left Affectiva and our relationship was somewhat strained; Dave, Affectiva's CEO, was on to a new job; and my beloved father-in-law had died. I felt rudderless and lost.

Most of all, I felt like a massive failure. This was the first time that I'd encountered a problem that hard work and perseverance couldn't solve. I felt like I had let everybody down.

But despite the dire predictions of my parents and in-laws of all the horrible things that would befall us if Wael and I split up, the reality was anticlimactic. The earth didn't fall off its axis, no one in Boston cared that I was separated from my husband, work was going well, and slowly, things began to look up.

We rented a house in a Boston suburb, and by fall, Jana and Adam were enrolled in an excellent private school that they both loved. We fell into a pleasant routine. I would drop the kids off at school in the morning, drive to Affectiva, and pick them up at the end of the school day. I no longer had to hide if I needed to catch up on email at night; there was no one judging me. On weekends, I'd drive the kids to their sporting meets and Jana to her harp class. Through the kids' school and work, slowly we began to build our social network in Boston. We also enjoyed family time; we'd take walks together and collect leaves in the autumn, and went on our first Halloween trick or treat. For our first Thanksgiving, I invited friends from Affectiva and added some Egyptian dishes to the traditional Thanksgiving dinner. It was a great relief to see how quickly the kids adapted to their lives here, yet they could still retain their cultural identity.

Nine months later, Wael and I decided to file for divorce.

The divorce awakened in me the ambition that had lain dormant during my grounded-in-Cairo year. I felt a new sense of urgency in driving Affectiva's success. "My children's livelihood depends on it," I wrote in my journal.

18

Woman in Charge

t was the beginning of the most exciting time in the AI industry. Half the U.S. population now owned smartphones, with 24/7 access to the Internet; texting was now the primary method of communication for Millennials and their younger siblings. Ubiquitous smartphone cameras had turned everyone into photographers; Instagram, Snapchat, Twitter, and Facebook had created new platforms for self-expression. And in March 2014, at the Academy Awards, host Ellen DeGeneres made history when she and a bunch of A-list celebrities took the selfie that nearly crashed Twitter, further fueling the growing selfie craze. Faces were in, but face-to-face, well, not so much.

The world that I had envisioned back in my grad school days at Cambridge University had arrived. The cyber world was merging rapidly with the real world. The lines were blurring, and we were now online all the time. But despite it all, our computers were still emotion blind. I had thought that by now we would be further along in bringing EQ to the digital world.

Under Nick's leadership, Affectiva had become a solid advertising tech company, but that was it. To me, this was way too limited a vision for the company. But I couldn't be angry with Nick: If I was unhappy with the way things were going, I had no one to blame but myself. I hadn't stepped up when I had the chance.

If there was a time for Emotion AI and Affectiva to break out, it was now. If we didn't, I feared that the industry would evolve around us, and we would lose our leadership role.

Once at the forefront of AI, Affectiva was now at risk of falling behind. Something new was disrupting our industry: "deep learning," a subset of machine learning. Although the terms are thrown around interchangeably, there is a significant difference. And it is one that could make or break a data company like Affectiva.

I think of machine learning as an assembly line with, in our case, the "product" being facial expression and emotion classifiers; the technical term is *machine learning platform* or *infrastructure* (as in *pipeline*). Keeping the process efficient and streamlined is as important as the product itself. You want to keep things flowing smoothly.

There are typical steps in the machine learning "assembly line," requiring different types of expertise. Data is now readily available, but it has to be collected; this is the job of the data acquisition specialist. In our case, we get face videos or audio files.

Second is data ingestion: These terabytes of data need to be stored somewhere, most often in the cloud. Data engineers are responsible for that.

Third is data annotation: These videos and audio files are not useful unless they are labeled or annotated by human experts—labelers. The labeler's job is to mark when important events happen, like a smile or a smirk or a frown.

Fourth, machine learning scientists are the ones who build the algorithm, as I did when I was working on my PhD at Cambridge. The machine learning scientists team up with quality assurance engineers to test the accuracy of the models. This involves quantitative testing on hundreds of thousands of examples as well as qualitative testing. Our quality assurance engineers will spend hours in front of a camera making faces, trying to break the algorithm.

As an AI company, our goal is to minimize the number of times we have to go through this cycle. If, for example, you have to repeat the cycle five times to get to an accurate smile classifier, versus, say, doing it in one iteration, it costs the company time and money and results in slower time to market.

The machine learning approach that we were using at the time at Affectiva was called "feature engineering." The machine learning scientists—most have a background like mine—would show the algorithm what to look for: For instance, in the case of training a smile algorithm, you would direct the algorithm to look at the mouth region, specifically, the movement of the lip corners. But for a raised eyebrow, you would direct it to focus on the eyebrows.

In 2015, I was advocating to switch out the feature engineering and replace it with deep learning, or deep neural networks, which enable the algorithm to figure out where it needs to focus just by observing lots and lots of examples of a smile (and comparing it to lots and lots of examples of non-smiles). This deep learning not only speeds up the process but, as is clear from the academic literature, gets you to more accurate classifiers faster, which means less of these iterations, which eventually means faster time to market.

I knew that if we were to remain an innovator, it was essential that we switch our technology to deep learning. But Nick resisted. He felt that we should devote our time and energy to building new emotion algorithms that would expand our repertoire of emotions and pay off immediately.

Switching over to deep learning was a big undertaking that would require changing the underlying algorithm. It was also labor-intensive and time-consuming. It would be a major project for a start-up like Affectiva. It would take six months and the full attention of two people on my team to move us into deep learning. I know this for a fact because I did it behind Nick's back. I took two of my machine learning scientists aside and told them, "This is a secret project. I'm giving you the green light. Don't tell anyone else in the company. When it's done, we'll tell them that we did it." It didn't take any convincing on my part; they understood the importance of what we were doing and were thrilled. Those on the computer science end of the business understood that deep learning was the future. If we didn't get on board, others would leave us in their dust.

The switch to deep learning paid off big-time. When I was contemplating the switch, Affectiva had five basic facial expressions: smile, brow furrow, raised eyebrow, disgust, and smirk. Since the transition to deep learning, we now have twenty facial expressions, six emotional states, and we are moving rapidly to complex cognitive states like fatigue and drowsiness. All this was made possible by deep learning.

I may have been maneuvering behind Nick's back, but at least I was no longer "grounded in Cairo." And I took advantage of my newfound freedom as Affectiva's founder and chief technology officer to accept speaking engagements at industry conferences, the kinds of opportunities I had had to turn down before. In January 2015, I received an invitation to speak at the TEDWomen event in Monterey, California. I had been recommended by the chief marketing officer of GE, Beth Comstock, who had seen me speak in Dubai that past fall.

TED gives its speakers access to the top thought leaders in the world, as well as an opportunity to deliver your message to millions of potential online viewers. A TED Talk is only twenty minutes long, but it takes hours and hours to prepare. They are well choreographed and planned. My primary contact was June Cohen, who served as producer of the event. I was assigned a group of about a half-dozen coaches to help me prep my talk and slide presentation. Each coach viewed my work through a different lens. One coach in particular, Dale Deletis, worked on my delivery, encouraging me to think of my presentation as a gift to my audience. That clicked with me because it made me less self-conscious and more focused on inspiring and intriguing others.

I took the talk *very* seriously: My coaches and I rehearsed end-lessly, and they provided detailed feedback. By the time I got to Monterey, I had practiced my presentation dozens of times. I was still reviewing it the night before in my hotel room when my daughter, Jana, sent me a text message—in all caps. Now in seventh grade and on her school's speech and debate team, she was already a better speaker than I was. The text read, "GOOD LUCK MOM. DON'T PLAY WITH YOUR HAIR." (Okay, I'll admit it—I often do this when I'm nervous.)

Jana's message cracked me up. More to the point, her text was exactly the argument I intended to make in my talk. People's primary method of communication was now text in an emotion-blind cyber world. I felt that Jana's text was so fitting that I emailed the TED team that evening and asked them to insert a new slide. TED frowns on changing anything at the last minute, but the organizers were so amused by Jana's text that they agreed to include it.

The next day, I walked onto the iconic TED stage, looked out at all the women (and a handful of men), and felt at ease. I owed a debt of gratitude to my TED team of professional coaches, but I realized that my training had begun long before, with a father who encouraged a five-year-old girl to stand on a chair and speak up. "Emotion," I began, "influences every aspect of our lives. From what decisions we make, to how we connect and communicate." When I got to the slide with Jana's text message, everyone in the audience laughed.

In preparing for the TED Talk, I had to really hone my message. First, I was encouraged by the team to share my "origin" story, the formative moment that had inspired my research. Reluctantly at first, I spoke about being a young married Egyptian woman moving to Cambridge, finding herself alone with her laptop as her only companion. My coach made me understand that I wasn't alone in what I had experienced, that my story would resonate with that of countless others. A lot of us feel disconnected when we're on our devices.

Perhaps most important, my coaches urged me to think big, to focus not just on what my company was doing now, but on its potential for the future: how this technology could transform people's lives. In my talk, I never even mentioned our work in market research. Instead, I talked about Affectiva's potential impact on mental health, autism, education, enhancing human relationships, and robotics. I spoke of how this technology would not only transform human-to-computer interfaces, but would fundamentally change human-to-human connection. I focused on the applications that resonated with people and touched something inside me. As I spoke, I knew that, more than anything, I wanted to make this new world a reality.

Practically overnight, I was elevated from a computer scientist and entrepreneur to a thought leader, a spokesperson for the future of this new form of AI. But when I returned to the office from the TED weekend, revved up and excited, and looked around at the reality of our work, I felt let down. The company seemed quiet, even moribund, to me. The sizzle was gone. I was angry with myself, and with Nick. This was not why I had started Affectiva. This was not why I had jeopardized my family. I had made so many sacrifices for Affectiva, but we were just going through the motions, treading water.

I understood the problem now: I needed to be in control. I had to be CEO.

My TED Talk was posted online in June and quickly received more than a million hits. For the first time, I was speaking openly about adding emotional intelligence to our machines, and the public found it exciting. People were more than receptive. They were eager for that capability. And they inspired me to move forward more aggressively.

First, we needed to forge our identity as a leader in this new kind of AI. When we first spun out, I had the opportunity to meet Todd Dagres of Spark Capital, one of the leading venture capitalists in the Boston area. I will never forget the advice he gave me: The best companies, he said, are those that define a new category, name it, seed it, and lead it. There are numerous examples of companies that have done this very successfully—for example, Uber and "ride sharing," Facebook in terms of social media, and Venmo for micro transactions with your friends.

It was time for Affectiva to define its category.

Gabi Zijderveld, our chief marketing officer, and I started brainstorming how we could best summarize what we did. We were recognized as an AI company, but we realized that there was no category yet for our kind of AI. We had to create one to make clear to potential clients, partners, and investors where we fit in. We talked about how we build our software, its applications, and the ethical implications of what we were doing. Emotion recognition? Emotion analytics? Emotion sensing? All these only partially encapsulated what we were trying to build. We particularly liked *artificial emotional intelligence* because it underscored that machines, like humans, need emotional intelligence to reach their full potential. But the term was awfully long, even cumbersome. So, we shortened it to *Emotion AI*. That same day, we stuck #EmotionAI into a tweet. We began a strategy to evangelize the concept, talking about it during press interviews, incorporating it into my keynotes, and using it in social media. We painted a vision of a world that didn't exist yet, but in doing so, got people excited about the prospect of a more human-centric technology. Eventually, it took off.

Suddenly, Affectiva was bombarded with countless requests from investors, scientists, and executives across several industries. A researcher in Boston wanted to create an app to predict and ultimately prevent suicide. An online education company wanted to use our technology to monitor student engagement and predict learning outcomes. A human resources company saw our technology as the perfect tool for screening new hires. A group of entrepreneurs wondered if we could create emotion-aware classrooms. A professional career coaching company thought it would be an excellent training tool. One great potential partner after another reached out to us.

I was determined to make all this happen. Brimming with ideas about how to inject new energy and excitement into Affectiva, I wanted to rally the team, to evangelize them behind the importance of bringing EQ to our computers.

That year was a turning point in my life: I emerged as a leader in my field. I received the 2015 Ingenuity Award from *Smithsonian,* the official journal of the Smithsonian Institution in Washington, D.C. The award recognizes the "shining achievements, of individuals in nine categories, who have had a revolutionary effect on how we perceive the world and how we live in it." I won for technology. (The following year's winner in that category was Amazon's Jeff Bezos!)

The award was presented at the National Portrait Gallery, a historic art museum in Washington, D.C., with an extraordinary collection of American faces from Colonial times to the present. Jana and Adam attended the awards ceremony along with my mom. We all dressed up for the occasion; it was a special night. I was in the process of becoming an American citizen, and when I stood up in front of the crowd to accept this award, I felt the magnitude of that privilege. The United States honors and celebrates innovation and entrepreneurship in a way that would have been very unlikely back in Egypt, and I felt a responsibility to make the most of this opportunity, to share the goodness and pay it forward.

More than ever, I really wanted to be at the helm of Affectiva, moving it in a meaningful direction. In January 2016, I shared my ideas and challenges with a mentor of mine, someone I often turned to when I needed career guidance or to work through a problem. After describing my situation at Affectiva, he confirmed my own growing conviction: "Rana, you need to become CEO."

I shook my head. "Nick is CEO, and he's not going anywhere." My mentor encouraged me to visualize becoming CEO, to visualize the path that I would need to take to get there. I knew I wasn't the sort of person to stage a coup, to try to turn the company's board against Nick the way John Scully had with Steve Jobs at Apple decades before. That wasn't my style.

A few days later, I was chatting with our VP of engineering, Tim Peacock. Out of the blue, apropos of nothing, he said, "You know, if you ever become CEO of a company, I'd like to be your COO." I was floored. I thought highly of Tim. I respected his experience and opinions; his support meant a great deal to me. Maybe it was time to make my move.

First, I had to negotiate with the voice in my head telling me, "Don't do it, Rana. You've never been CEO. You will fail and bring the company down with you!" That was the same voice that had kept me from stepping up in the first place a few years earlier, and now I had nothing but regrets. I decided to prove to it (to me) that that voice was wrong. So I set out to build a case in my favor. In retrospect I realize I was my toughest critic; it was harder to convince myself than it was to convince others.

What does a CEO do? I asked myself. I began to make some notes in my journal:

A CEO "evangelizes the vision/mission of the company to internal and external stakeholders." I was already doing that.

A CEO "sets the company and product strategy including IP/science roadmap"—just as I was trying to do with deep learning.

A technical founder in the CEO position "is attractive to investors as well as is a talent magnet." Deep technical expertise gives investors confidence, and AI scientists want to work for an industry leader.

A CEO must raise money—something I was already immersed in

The more I studied, the more I realized that I was already handling many of the duties of a CEO; I just didn't have the title. *If I can't be the CEO*, I thought, *why not become co-CEO?* I ran the idea by a few colleagues privately, including several who had been co-CEOs themselves, and every one of them said the same thing: "Bad idea. Too much conflict and confusion for the team. Don't do it."

But I couldn't think of another way forward.

By March, I had summoned my courage and went in to talk to Nick, making what I thought was a strong case for him to make me co-CEO.

Nick looked surprised at my suggestion. He obviously didn't see things the way I did. At first, he flatly refused my suggestion. And the more we discussed it, the more I realized that I actually didn't want to be co-CEO. I wanted to be *the* CEO.

As Nick and I continued to talk, fate intervened. A major tech company's chief technical officer reached out to me with a *very* lucrative offer. They wanted to hire me, badly, but weren't ready to acquire Affectiva. To me, this was tantamount to getting a marriage proposal with the caveat "I love you, Rana, but I don't want your kids."

I knew, however, that if I were to stay at Affectiva, I had to be in control. So, I used my offer as leverage. One warm day in March, before I left Boston to give a talk at an industry conference in New York, I gave Nick an ultimatum: "Either I become CEO of Affectiva, or I take the job offer" (which would have been financially more lucrative for my family anyhow).

My heart was beating fast. I was nervous beyond words, although my training in emotion recognition helped me mask that nervousness.

Realizing I was serious, Nick promised to think about it.

While I was on the train back to Boston that evening, Nick called me. After giving it a lot of thought, he said, he realized that Affectiva was my baby. I was passionate about the company in a way that nobody else could be. He graciously agreed to resign, step into the role of chairman for a while to ensure a smooth transition, and support my pitch to the board.

At that moment, I was so terrified at being given what I had desperately wanted that I almost told him, "Just kidding. Never mind!" But I refused to let my nerves get in the way.

A few weeks later, Nick and I went to the board and explained our thinking. The board took a vote, and on May 12, 2016, I was appointed CEO. Nick stayed on as chairman, a generous act. Looking back, I see that the hardest person to convince that I could be CEO was myself.

On the day I became CEO, the first thing that I did was to email

Roz and let her know that I was now CEO, that I still believed in our original vision and mission, and that I would do everything in my power to achieve it. Roz responded with a gracious congratulatory email. Later, I gathered the team together and shared with them my vision for the company. Affectiva, I told them, defined the Emotion AI space. "We own it," I said, and now it is time to forge full speed ahead and grow Affectiva with a new round of capital. "It's an exciting time to be in AI," I said. "We have the potential to transform many industries."

We are problem solvers, I told them. We solve problems through our collective intelligence. Every person on our team has the ability to have a huge impact on our direction, strategy, and products. I told them that I wanted our employees to have autonomy, to take initiative, and to make things happen.

Now that I was at the helm, I savored the opportunity to reset Affectiva's culture to create new energy and excitement at the company—the kind of excitement I had felt at the MIT Media Lab. I therefore instituted a weekly meeting for the entire company. Any question was fair game, and all ideas were welcome. I knew that we needed to swim in a bigger pond. One of my first tasks was to move the company to Boston, to be part of the fast-growing Boston tech scene. We had become too inwardly focused. I wanted us to be far more collaborative, to expand our partnerships and bring in young talent with fresh ideas.

The night I became CEO, I went home and wrote the following in my journal:

Great day. I can't believe this is happening. Owe it to Nick big time.

I must remember that my job is to make these people successful and stars in their own circles.

I need to work on maintaining the right work/life balance, and spend time with my kids. Can't make the same mistake again.

Tomorrow is Uncle Ahmed's birthday—how I miss him! He

believed in me; he told me that God gave me a gift that made people want to rally around and help me. What a powerful gift: The ability to galvanize people and lead them to change.

It's a responsibility, too, and one that I take very seriously now that I am CEO.

Ok . . . 3:30 am. Need to sleep.

PART IV

A Pioneer in AI

19

Hacking the Hackathon

f I've learned anything from launching and running a start-up, it's that you need to focus. You can't be all things to all people; you must figure out who you are, what you're good at, and where you fit in the marketplace. Yet the potential for Emotion AI is so vast, with so many possible applications, that I was frustrated by the fact that Affectiva had numerous requests from people and organizations who wanted to collaborate with us on worthwhile projects in nearly every field imaginable. We didn't have the staff or resources to explore all these potential uses and many of these people didn't have the money to pay our licensing fees. I felt that it was unfair to hold back technology from people who could do so much good with it. So we hosted our first hackathon, Emotion Lab '16, where we made our software available to a diverse group of participants who were given free rein to use it any way they liked.

A hackathon is a lot like a TV celebrity chef cook-off, where the contestants, limited to a few ingredients, whip up a fabulous meal within a tight time frame. Similarly, in a hackathon, participants with just a few key technology tools at their disposal must transform their idea into a working prototype, usually over a weekend. Both competitions typically end with a panel awarding prizes to the top creations.

Hackathon is a blend of the words *hack* and *marathon*—and it is a race against time. A *hack* refers to tinkering with a system and altering it in one way or another, and that can be for the good.

Emotion Lab '16 was held at Microsoft's New England Research and Development Center, or NERD—no kidding!—a vast, airy workspace on the MIT campus. We kicked off the three-day event by giving everybody an opportunity to introduce themselves and describe what they did. When it was her turn, Kim (not her real name), a shy, bespectacled woman in her late thirties, announced that she was transgender. In the Middle East, topics like sexual orientation and gender identity are pretty much never discussed in private let alone in public, and I admired Kim's forthrightness. Later, I noticed that she was sitting alone on a couch. I sat down next to her, and we started to talk.

I learned that she held a PhD in chemistry from MIT. She admitted, however, that her academic success felt hollow because her family, and especially her parents, did not accept her as a female. Kim knew from an early age that, despite her anatomy, she identified as female. Eventually, she got tired of living in conflict with her true feelings, and that's when she transitioned. She felt good about her decision but was heartbroken that her parents simply couldn't accept her as a woman.

I had an idea. Our software has a gender classifier that identifies whether a subject is male or female based on how they look. I asked Kim, "Why don't we run an experiment? Why don't we see what gender the software says you are?" I cautioned her that the results could go either way. I wasn't sure what the algorithm would make of someone who had transitioned from male to female—the algorithm had not yet been trained on transgender people, so I didn't know how it would interpret Kim's face.

I turned on our app, and Kim looked into the camera on my laptop. The thirty seconds it took to calibrate felt like an hour. For a moment, I wondered if I had made a horrible mistake, if my experiment would turn into something hurtful. Then, seconds later, the female icon popped up on the screen, and it was also wearing glasses, just like Kim.

Kim high-fived me and smiled broadly. Her smile? It was rated 100 percent probability score, implying that it was a high intensity smile, but you didn't need a high-tech tool to see that she was truly happy. She then asked if we could take a screenshot of the female icon to send to her parents. She immediately followed up with a phone call and said, "See, science says I'm a female!"

I come from a culture where people like Kim are not accepted—they are "the other"—and that's the case in many cultures. This was a special moment for me: Kim and I connected as two human beings. In that instant, I felt enormous empathy for her. I understood her longing to be accepted by her family and by society, and I thought, *Isn't that what Emotion AI is all about?*

Emotion AI is about humanizing our technology to promote better understanding and better human-to-human connections. To do Emotion AI well, it is essential to include a broad swath of people. Emotion Lab was designed to be inclusive and draw from many different perspectives; that was one of the main reasons we held a hackathon in the first place!

Hackathons attract mostly programmers, usually men. So, we did a few hacks of our own on the typical hackathon. We didn't want to exclude anybody—male programmers were certainly welcome—but we also extended invitations to women's groups in the area, holding the first twenty slots open for women. We thereby achieved equal numbers of women and men, a rarity at these events. We strove for international diversity, too, and had representatives from all over the world: Sweden, England, Egypt, Japan, and Israel. And in what is a real radical departure from the status quo, we invited people from all walks of life. We reached out to people in other disciplines. We made sure that we had academic professors, artists, musicians, project managers, graphic designers, educators, autism researchers, psychologists, public health advocates, and the like. And we enabled these non-technologists to innovate side by side with computer pros, something that's not done enough.

We even invited one of the other start-ups in our space, Beyond Verbal, to participate. Beyond Verbal is a Tel Aviv–based company that specializes in vocal analytics. Most companies don't invite peer companies or potential competitors to their events, but we realized that making this technology available for our hackers would allow for more sophisticated projects.

Now, it is not enough just to extend invitations to groups that don't normally come to tech events. We had to make attendance possible also for people with family responsibilities, who couldn't disappear for days on end. And we had to make them feel welcome, something the tech community at large doesn't do well. For one thing, the round-the-clock nature of hackathons often means that it is difficult for mothers (and even fathers) to get away for a weekend. For another, the "bro" culture of men locked in a room gulping down Red Bull all night and eating pizza—well, that's off-putting to many of the people we reached out to. So, we decided to do things differently. We didn't work straight through. We shut down at night so that people could go home. I'm a big advocate of people getting enough sleep. And because so many employees in our company have kids, during the day we offered a supervised, parallel program for children in which they could build their own projects.

Did any of these "hacks" to the typical hackathon matter? I think so. Of course, like all competitions of this type, things got intense. The teams worked super hard. But everyone was made to feel like they belonged.

Each participant had an opportunity to pitch their project to the crowd to see if they could build a team. Ultimately, ten projects were selected, and the sixty participants divided up into teams. Each team had access to the same "ingredients" to bring their vision to fruition: Affectiva's Emotion AI; Beyond Verbal's software; Pavlok's wearable wrist sensor; Brain Power's Google Glass; Jibo the robot; the *Star Wars* franchise's BB-8 toy droid; and Arduino, an open-source electronics platform. The only requirement was that Affectiva's Emotion AI be integrated into the prototype. Other than that, there were no rules.

Not only were the groups diverse in every way, but so were the projects. A videogame with an unlikely name, *Murderous Llamas*, used facial expressions (instead of controllers) to navigate the players. The app Blind Emotion Aid hacked Google Glass to enable the visually impaired to "see" the emotional state of the people with whom they were interacting. "Super TA" transformed the BB-8 droid toy as a tool to provide real-time feedback for teachers to assess student comprehension and attention. These are all new and exciting uses for Emotion AI, and they came about because of the unique mix of participants.

One project tackled a societal problem that few people want to talk about—suicide. When Steven Vannoy, PhD, MPH, associate professor of counseling and school psychology, College of Education and Human Development at the University of Massachusetts, Boston, made his pitch to build a suicide prevention app, it was received with enthusiasm by the group. Two techies stepped up to work with him.

In the 1980s, Vannoy was a computer programmer, but a decade later he decided he wanted to pivot into a more people-oriented profession. So, he switched to psychology, earning a master's in public health and a postdoctoral fellowship in geriatric mental health services, with a specialty in suicide prevention.

Suicide is on the rise in the United States and the world. In 2017, the latest year for which figures are available, 47,173 people died by suicide in the United States and another 1.4 million people attempted suicide. According to the U.S. Centers for Disease Control and Prevention (CDC), 5 million more have suicidal thoughts. Even the most skilled clinicians can't predict who among their patients will carry through on thoughts of suicide and who will not.

Teenagers and young adults are especially vulnerable: Suicide is the second leading cause of death among fifteen- to twenty-four-year-olds. Vannoy says that among the students with mental health issues he recruits into his research projects on campus, "we find a significant portion with an elevated risk for suicide even when we're not selecting for it."

A major problem is that, as a field, mental health has not embraced the use of new technologies to assess and track patients. Despite the fact that medicine has become automated and AI is being implemented to both diagnose and treat serious illnesses, mental health practitioners have pretty much steered clear of tech. Patient assessments rely primarily on self-reporting, which can be skewed. For example, a doctor may ask a patient, "Are you feeling suicidal?" or "Are you depressed?" and the patient may or may not respond honestly due to the stigma associated with mental health issues or may be highly ambivalent, leaning toward life at that particular moment, but hours later leaning toward death.

Of course, there are many people who are depressed or anxious and at risk for suicide who have not reached out for help. According to the World Health Organization, two-thirds of people worldwide who have mental health issues don't seek help from a mental health professional. But even if a patient at high risk for suicide is seeing a practitioner, in all likelihood it's just for fifty minutes a week. Such patients are left to their own devices the rest of the time.

Given his technology background, Vannoy wondered why mental health was lagging behind other medical disciplines, but he didn't know how to remedy it. He had been out of the computer programming field for nearly two decades, and much had changed. When he read an article about Emotion AI in *The New Yorker* that mentioned Affectiva, he saw a way to bridge the two disciplines. He wondered if an Emotion AI app for a smartphone could be used to track patients in real time, in the real world, so that a specialist could intervene if it appeared that a patient was severely depressed or about to inflict harm on himself. Vannoy contacted us with his idea, and we invited him to the hackathon, where it became a reality.

Vannoy and his team called their prototype, a suicide prevention app for a smartphone, Feel4Life. It uses Affectiva's Emotion AI and Beyond Verbal's voice software to pick up signs of distress or a change in mood in the user. Unlike standard assessment tests, the app doesn't ask if the person is feeling depressed or contemplating suicide. Instead, it tries to draw him out to talk about what is happening in his life, asking questions like "How's your day going?" "What are you thinking about?" "What are your plans for the rest of the week?"

During every check-in, the user's responses get compared to previous answers to see if they're more positive or negative than their baseline and their most recent check-in. The app was designed to conduct three check-ins per day, and the user's provider will be able to see this log. If the user isn't checking in or if he sounds "off," the provider will be notified and will intervene and/or alert a family member. Feel4Life is just a prototype and is a long way off from being developed, but if it ever made it to market, Vannoy stresses that it would not replace the fifty-minute weekly therapy session. Rather, it would augment and extend the practitioner's reach.

Beyond asking questions, tracking the online behavior of people at high risk for suicide could also produce important information that has previously gone undetected, Vannoy notes. For example, it could detect whether people who are truly suicidal are drawn to different types of news and information online compared with those who are not; or whether they pay special attention to stories about suicide or death.

"It's possible that they may be excluding in their world information about the future and hope," Vannoy notes. "It may turn out that people who have a history of suicide-related behavior and who are thinking about that may spend a lot more time looking at death-related images and self-harm-related images than things that are related to the future and human connection. We don't know the answer to that. That's what we're trying to find out." Certainly, automating this kind of research could help gather much more data and, ultimately, save lives.

In the beginning, an app like Feel4Life could be targeted to people whom we already know to be at very high risk for suicide. But the real power of this technology is to integrate it into the technology we use every day, like Alexa or Siri or Cortana. Depression is common in our society. According to the American Psychiatric Association, one in six people experiences depression at some time in their life. Often, people don't even realize when it's happening to them or a family member. Monitoring something as sensitive as mental health needs to be done only with the individual's consent, and the data must be kept private. In the case of early detection for suicide, an app monitoring it could save thousands of lives each year.

If it hadn't been for the hackathon, it is unlikely that the Feel-4Life app would have made it to prototype or, for that matter, that any of the projects completed that weekend would have seen the light of day.

When we held Emotion Lab '16, we wanted to see what a group of creative, passionate people could do with Emotion AI. By the end of the weekend, we had a glimpse of what an Emotion AI world would look like. Such a world is compassionate, fun, useful, thoughtful, and unafraid to tackle difficult issues. It is *empathetic*, aware of the feelings of others and sensitive to their needs.

Ultimately, Emotion Lab turned out to be a prototype of its own, a new way to "do" technology, to make it responsive to the needs of both makers and users, to open the field up to people who thought there was no place for them in this world.

We need *everybody* to come together in designing and deploying these AI systems. If we continue on the Silicon Valley tech track, where only a very small number of people design systems for the rest of us, technology is going to be biased—perhaps not intentionally, but we will be unwittingly duplicating the biases that exist in society.

In designing the systems—from the conception of an idea to the collection of the actual data, to the machine learning algorithm, to the deployment—we need diversity, across the board. We need *everybody* to be at the table: Our technology has to detect the smiles of people living in remote areas of India, the smiles of hijabi women, and the smiles of transgender people. Our algorithms must work on people from all walks of life, of all complexions, genders, ages, and ethnic groups. And our algorithms must be used to help improve human life and solve some of the long-standing problems that are hurting us as a society and diminishing the quality of life for people whose needs have been overlooked.

20

Gone Quiet

My colleague Emily first noticed that her two-year-old son, Matt, had "gone quiet" during a playdate with his two cousins, who were just a couple of months older. (These names have been changed in order to disguise the individuals' real identities.)

"His cousins were talking away, and Matt wasn't uttering a word," Emily recalls, which seemed out of character because he had been an early talker. Now he was silent most of the time.

Emily began to observe other behaviors in Matt that raised some red flags. He rarely made eye contact, and when he did, he seemed uncomfortable and quickly looked away. He didn't like being hugged. When she called his name, nine out of ten times he wouldn't respond. And when Matt played with a toy car, he didn't race it around the floor, like most boys his age do. Instead, he turned it upside down and fixated on the wheels.

Emily is not an expert in child development, but she knew enough from what she had read in parenting books that these behaviors were often associated with autism. When she raised her concerns with Matt's pediatrician, the doctor assured her that wasn't the case. After all, Matt's behavior was hardly that of the stereotypical autistic child. Although he was shy, he didn't "stim" (i.e., engage in self-stimulating, such as rocking or other repetitive movements) or walk on his tip-toes (which toddlers learning to walk often do, but most outgrow it). In other words, he didn't exhibit any of the obvious signs of autism.

"It's just a language delay," the pediatrician reassured her. "All he needs is some speech therapy."

Emily took Matt to see two additional child psychologists, who agreed that he was fine, and denied her request to have him enrolled in a state-funded early intervention program for autism. The out-of-pocket costs would have been prohibitively expensive.

"They looked me in the eye and said, 'Ma'am, we see ten kids a day who are on the spectrum, and your son definitely is not,'" Emily says.

She wanted to believe them, but she kept seeing behaviors in Matt that alarmed her. The scientist in her couldn't let it go, and the more she learned about autism, the more she worried about Matt's future. Although there is no cure for autism spectrum disorder, Emily had read research studies that showed a strong correlation between being diagnosed before age three and positive outcomes in therapy. If that was true, then the window of opportunity was closing.

Determined to get to the bottom of her son's condition, Emily found one of the leading authorities on autism in her state and put her name on a four-month waiting list for an appointment. Just a few months shy of his third birthday, Emily took Matt in for an evaluation. Within a few minutes of interacting with him, the autism specialist told Emily, "You're right. Your son is on the autism spectrum."

Emily reached out to me and asked if there was any technology I knew of that could help Matt. You remember that crazy idea that Roz and I had to build a Google Glass–type device that could decipher facial expressions for autistic kids? It was now a reality, but we weren't the ones developing it.

THE EMPOWERED BRAIN

In 2013, entrepreneur Ned Sahin, a Harvard PhD in neuroscience with a master's in cognitive neuroscience from MIT, was attending an all-day autism seminar at MIT. He knew very little about autism and was surprised by "how very far behind we were on autism relative to other conditions, especially considering how many people are affected."

From these researchers and members of the autism community, Sahin heard about the struggles first-hand confronting parents like Emily who take their children from doctor to doctor before getting a definitive diagnosis. Young adults on the autism spectrum spoke about their difficulties obtaining employment and maintaining relationships. Parents of noncommunicative children who have severe forms of autism described how they desperately wanted to understand what their children are thinking and feeling.

"What struck me the most about the meeting wasn't the science; it was the short preamble each presenter gave to justify why they were working on this problem," Sahin recalls. "It was the *human struggle* that captivated me."

And what was most surprising to Sahin, despite the growing prevalence of the problem—one in fifty-nine children in the United States has been diagnosed with autism spectrum disorder—was that there were not enough specialists in the States or elsewhere to treat autism or to support caregivers.

Sahin was primed for a new challenge. He had just taken a year off from his career to travel to more than two dozen countries with his wife, Nicole. Although he and his wife left their cellphones at home—they wanted to interact with people and not Google Maps—

Sahin noticed how ubiquitous smartphones had become, even in remote areas of the world.

"I had an epiphany on that trip," he says. "We saw the different challenges confronting people in many of the countries we visited, and how people struggled in their everyday lives. I realized that software has global reach, will rapidly scale to meet demand, and can really help level the playing field for people who need support."

Sahin's return to the States in 2013 coincided with the launch of Google Glass, smart glasses with a built-in camera and a head-mounted optical display. Google Glass, with its high price tag and hyped marketing, had its share of critics—at the time, those who wore them were called "glassholes." But Sahin saw Glass as a transformative tool for helping people with neurological disorders.

But the Glass was still hard to come by. Not only was it expensive, but it was being released very slowly, to a select few. So, Dr. Sahin went to Google's headquarters in Mountain View, California, and, after some skillful maneuvering through the company hierarchy, got Google to donate a Glass so that he could hack it for his own purposes.

Once Sahin had the Glass in hand, he kept brainstorming uses for the new technology. But nothing seemed quite right—until the conference on autism. Then everything seemed to click. Now Sahin knew exactly what he was going to do with his Google Glass.

"Many autistic people are not cognitively impaired, but they are socially impaired, and that's what was getting in their way," he said. "I thought that if we combined AI with a wearable computer, we could develop an outsourced *social interface* with other people and use it as a training tool so that, eventually, users can reach their full potential."

In other words, Sahin wanted to build a Google Glass that decoded facial expressions and other social cues to teach autistic people to interact better with others. Just like Roz and I did seven years earlier, when we applied for the National Science Foundation grant but were turned down because the grant reviewers decided it would be impossible to build. Seven years is a lifetime in the world of tech. AI had leapt forward, and the sleek Google Glass was light-years ahead of our clumsy device. That year, Sahin launched his start-up, Brain Power, to unlock the power of the brain (specifically the autistic brain) with neuroscience-based software and hardware that transforms wearable computers into neuro-assistive educational devices that are engaging and encourage human interaction.

Sahin brought in a coder from MIT and started building a staff of neuroscientists and technologists. He made it a point to hire people on the autism spectrum, on whom he relied for advice on and insight into the condition. Sahin and his team consulted with family members and caregivers of autistic children and adults, as well as with the individuals themselves, for help in designing the technology. Ultimately, it resulted in a suite of apps that teach social-emotional skills and encourage users to be engaged in the world around them.

None of this was easy. As Sahin discovered, Google Glass proved difficult to hack. It wasn't built like a user-friendly smartphone compatible with downloading apps. It was mind-bogglingly complicated to make the changes he wanted, even for highly skilled technologists. Eventually, however, he and his team made it work.

His efforts bring back memories of my days at the MIT Media Lab, cobbling together a primitive version of a wearable tool for autism. I feel a pang of nostalgia when I think about how dedicated I was to building the iSET back at MIT. It was hard to let it go when we started Affectiva, but we had enough on our plate running a company.

When Sahin began his work, he had no idea that Roz and I had undertaken a similar project, and we knew nothing about what he was doing. Given the MIT connection, however, it was inevitable that our paths would eventually cross. He found Roz, and Roz put him in touch with me.

We met over coffee and became fast friends. I understood his mission and saw his work as a means to fulfill our initial vision in a way we never could have done on our own. So, even though he couldn't afford to pay our licensing fee—his company wouldn't be making money for some time—we gave him free access to our software, and even threw in a few hours of consulting every month. It was a big ask for my team, who can license out our software for significant sums of money, but we do this kind of thing because it is the right thing to do. I remember how I benefited from the generosity of Simon Baron-Cohen, who gave me access to his database when I was pursing my PhD. It helped move my work forward by years, and when Brain Power needed my help, I was in a position to pay it forward. Ultimately, sharing technology with people like Ned is good for business as well as company morale. It helps us attract the very best employees, especially among Millennials, who value companies that have a purpose and are mission-driven.

Today, Brain Power's product, the Empowered Brain, is a suite of game-like apps that enables children and adults with autism, attention deficit hyperactivity disorder (ADHD), and other brain-related challenges to learn social-emotional skills. The apps run on Google Glass and employ augmented reality and Affectiva's software to coach users to engage in conversation and decode the facial expressions and emotional states of the people with whom they are interacting. The software also pairs with a smartphone or tablet, and features an account dashboard that displays hundreds of data points that Glass records. The data, which is cataloged in real time, lets parents and teachers know how the user is progressing.

To Sahin, the wearable headset is a critical part of the learning process. He points out that an app or tablet wouldn't work as well because it encourages people to look down. "If you're looking down at a screen, your sensory input from the world is tightly controlled," he notes. "If I take an iPad away from you, your head automatically goes up and your ears are pointing toward a different part of the world, which, for a child on the autism spectrum, can be overwhelming and disruptive."

In contrast, when using smart glasses, the user is "heads up, hands free, and engaged" with the social world around them.

"The technology is a scorekeeper and teacher for human-to-human interaction," Sahin explains. "To win points, you have to literally interact with another human being in the real world."

Each game offered by Empowered Brain is designed to enhance human-to-human interaction. Take Emotion Charades, for instance: Imagine wearing Glass while interacting with a friend. To accrue points (in this case virtual gems), you have to guess what emotion your partner is depicting by tilting your head to select one of two options; like a happy emoji or a sad emoji. So, if your friend is smiling and you pick the happy emoji, your friend would reward you with virtual gems. But if you incorrectly identify your friend's smile as sadness, and pick the sad face emoji instead, you get no gems. The game teaches you how to identify expressions of emotion during an interaction and encourages users to make face-directed gaze during conversation.

The ultimate goal is to wean the users off Empowered Brain as they progress, so they can function more fully in the real world. For example, in the Face2Face game that promotes face-directed gaze during conversation, as the user advances in difficulty, some visual cues disappear.

Five years after Sahin launched his company, Emily reached out to me and asked if I knew of any technology that could help Matt. Instead of saying, "Well, maybe one day we'll have something," I was able to refer her to Sahin. A few months after Matt's diagnosis, he visited the Brain Power office in Cambridge, Massachusetts, and tried out the Empowered Brain system. In a way, although Matt is young for this technology, he is the perfect candidate. He is bright, and is doing well in academics in his preschool. He has caught up in terms of language skills, is good with numbers, and *loves* technology. But his social skills need work, and his inability to look at people directly and focus when he is speaking to them will hold him back as he forms friendships, moves through school, and eventually becomes an adult.

Empowered Brain helps kids like Matt deal with practical problems that impact everyday social interactions. The system shows Matt how to move his head in a way that is not distracting or off-putting to other people, so that he can maintain a conversation. And even if he avoids direct eye contact, he is at least looking toward the person and appearing to be engaged. With the help of Empowered Brain, Matt is learning how to carry on a conversation with someone else without looking away or down, and that is a major milestone for him.

Although it has proved to be successful across a broad age range, Empowered Brain was originally designed for school-age kids. So, when Matt began using it, the question was, would he, at his young age, understand what to do? But when he put on the glasses, he just got it, Emily says. "He was looking at an image of me on a screen, and every time he looked at my face or eyes, he earned points. He quickly understood how to play the game."

Right now, 150 Empowered Brain devices are in use, in more than a dozen countries. The company is currently focused on increasing usage in Massachusetts public schools and private homes with a goal to expand to other states, and some in Massachusetts public schools to assist special education students. With 550,000 autistic students enrolled in individualized education plans (IEPs) as part of special education programs throughout the United States, that is a sizable market waiting to be tapped.

In addition to its teaching a child life skills, Sahin views the technology as a tool to foster better understanding between autistic children and their parents. "That parent is the only person who is going to defend this young human against the world. And if she doesn't feel she understands what's going on with her child, she can get dispirited. It's a bridge, really, that helps build human connection and ultimately, a relationship."

It is a bridge that has the potential to improve communication and understanding between people on different ends of the autism spectrum, whether they are living under the same roof, like Matt and his family; learning in the same classroom; or working together in an office. This innovation can benefit a whole generation of people who need an emotion prosthetic to help them better manage their interactions with others.

Affectiva's partnership with Brain Power is not a moneymaking proposition, and it may never be. That doesn't matter. To me, it is the fulfillment of the mission I began nearly two decades ago: to build an emotion prosthetic for people who needed an EQ boost; to help people like Matt fulfill their potential and remove the barriers that have been holding them back. It feels right to have passed the torch on to Sahin, who is putting this technology to the best possible use.

21

Secrets of a Smile

Our face is our message board. It tells the rest of the world how we feel.

—JOSEPH DUSSELDORP, MD, facial reconstructive
plastic surgeon, Sydney, Australia

My smile is my superpower, and so is yours. When I was a foreign PhD student at Cambridge, in England, my smile helped me break down walls, build bridges, and forge strong emotional ties with others. Today, I use my "I come in peace" smile at work when I'm popping my head into someone's office with a laundry list of requests, or when I am meeting prospective clients, because I know that emotions are contagious and smiles are irresistible. If I'm smiling while asking for something, it will be perceived entirely differently than if I'm frowning or brusque while making the same request. Smiles are so essential for human interaction that, even before we are born, we practice our smiles in our mother's womb. Yes, smiles are critical for human-to-human communication.

So, imagine losing your ability to smile. That's what happened to Susan (not her real name), an elementary school teacher in Sydney, Australia, who woke up one morning with a, well, *broken* face. While one side of her face still functioned, the other half was frozen. So, her smile was a strange, droopy version of its former self, a distorted half-smile that didn't express on the outside what she was feeling inside.

"My smile is my whole profession. It's how I engage my students," she tearfully told her doctor. When students started pulling away, asking, "What's wrong with your face?" she decided to leave the classroom and work in administration.

The condition that robbed Susan of her smile is called facial palsy, a neuromuscular disorder resulting in lopsided and awkward facial expressions. It can strike quickly, and is caused by any number of conditions, ranging from stroke to viruses. Sometimes it clears up on its own; sometimes it doesn't. In many cases, facial reconstructive surgery is required to make the face work again.

Susan is a patient of Dr. Joseph Dusseldorp, an Australian surgeon and clinical fellow at the Department of Facial Reconstructive Plastic Surgery at Massachusetts Eye and Ear, part of Harvard Medical School. Although the condition interfered with her ability to eat, speak, breathe through her nose, and close her eyes, what she cared about most was getting her smile back.

It's not just adults who suffer from this condition; young children can be born with a developmental problem called Mobius syndrome, and in this case, both sides of the face are paralyzed. These kids are unable to laugh. They may have a well-developed sense of humor and be normal in every other way, but their facial muscles simply don't move. It is as if they're wearing a mask.

I met Joe at Affectiva's first Emotion AI Summit, which we organized in 2017, a meeting of experts from around the world focused on the uses of Emotion AI, including in healthcare. (Since then, we've held a summit every year.) Dusseldorp was doing a fellowship at Harvard Medical School and was curious about how Emotion AI technology might apply to his work. The topic quickly turned to smiles.

Dusseldorp told me, whether patients are adults or children, when they seek him out for reconstructive surgery, their reasoning is "I just want my smile back." As Dusseldorp says, "If we can improve their smile just a little bit, it makes a massive change to their personality and their whole life." That's why 98 percent of the surgeries Dusseldorp performs on adults and children with facial palsy use "smile reanimation," a complex microsurgical procedure reorienting the nervous system that supplies the electrical activity that moves the muscles of the face. As he explains, "Basically, it's a lot like what an electrician does: Think of it as taking an 'extension cord' from the healthy side of the face and diverting it across to the damaged side."

Smile reanimation is an arduous, complicated procedure for both surgeon and patient, and recovery can take several months or more. Surgeons closely monitor patients postsurgically, to assess their progress, specifically whether the broken part of the face starts to move again and, over time, whether full symmetry has been restored to the smile. Even the slightest asymmetry can take you from a perfect smile to a smirk. The goal is to re-create a convincing smile that passes what Dusseldorp calls "the man on the street test." In other words, does it look authentic or would a passerby do a double take and wonder, *Hey, what's wrong with that face?*

Even if the operation is a success, and function is restored to the face, subtle things could be wrong with the smile that are difficult to quantify. "Suppose even though the smile muscle starts to work, and the corner of the mouth is moving, the smile does not tell me that the person is happy," Dusseldorp says. "Maybe it moves too far laterally or too far vertically, but there actually could be all sorts of things about it that are wrong, and the smile actually communicates the wrong message."

I knew from my own research that there are dozens of different types of smiles and that each conveys a different meaning—and not all of them express joy. Even the tiniest variation around the mouth can change the message. But what bothered Dusseldorp most was the lack of an *objective* tool to monitor a repaired smile over time, like those available to other medical specialties. After heart or kidney surgery, a surgeon can use any number of tests to check on a patient's progress. When it comes to facial surgery, though, there is no standardized test to assess function; surgeons have to rely on their own two eyes to track the patient through recovery.

Postsurgical care for smile reanimation involves follow-up visits during which the surgeon typically tries to elicit "spontaneous" smiles from the patient by telling jokes or showing funny videos. Dusseldorp was dissatisfied with the unscientific nature of the postsurgical assessment—the entire process seemed hit or miss—yet he couldn't figure out an alternative. The answer came to him during a presentation at the Emotion AI Summit, an annual meeting sponsored by Affectiva that includes representatives from a wide range of disciplines to explore potential applications for Emotion AI technology. Graham Page, now head of Affectiva's Media Analytics Business Unit, but then with Kantar Millward Brown (KMB, formerly Millward Brown), the advertising and branding company that had partnered with Affectiva, described how KMB uses our software to track and measure the facial expressions of consumers watching video commercials. Although on the surface, it had nothing to do with medicine, when Dusseldorp saw Page's demo, he immediately saw a link between KMB's work and his patients. *That's exactly what we reconstructive plastic surgeons are trying to do,* he thought. *Except we don't have these awesome tools!* He realized that the technology could be used to develop the objective measure of surgical outcomes he had been looking for. So he asked us if he could access the software, and we said yes. And we also recommended that he view humorous videos we knew were in the public domain that he could show his patients.

Dusseldorp finally had a tool that not only could measure and rate smiles, but would also eliminate much of the guesswork involved in evaluating postoperative patients.

One of the first things Dusseldorp did with our Emotion AI software was to investigate how the smiles of people with facial palsy were interpreted by others. In one study, he ran videos of the smiles of presurgery patients through our software. He discovered that their smiles were sending others the wrong message. Although the patients were trying to express joy, an asymmetric smile, for most of us, is associated with contempt. To add to this distortion, people with a form of partially recovered facial palsy, known as synkinesis, may wrinkle their noses involuntarily when they try to smile, which is interpreted as disgust.

After smile reanimation, Affectiva's Emotion AI detected significantly more joy and less negative emotion in the patients' new smiles, a vast improvement. Using our software, Dusseldorp can now rate the type and intensity of a smile postsurgery, just as we do in market research when we test advertisements. The beauty of this approach is that, in the future, when using an app that Dusseldorp is developing with our software developer kit (SDK), postoperative patients will be able to monitor their own smiles in real time, in their own homes, eliminating the need for office visits.

"We don't always get it right," Dusseldorp admits. "Once we've got a good understanding of how this application really works, we can home in on results that aren't perfect. For example, we may think we've done a technically sound operation, but when the patient smiles, the smile is not expressing 100 percent joy—maybe it's 50 percent. This tool enables us to go in and look en masse at the cases that fall into that category, and then to consider what we need to modify in our operation to create more natural smiles for these people."

I still find the whole thing rather mind-boggling: Who would have thought that a talk by an advertising executive would plant a novel idea in the mind of a plastic surgeon whose professional work was about helping people smile again? It was precisely the kind of cross-pollination of ideas I had hoped the summit would produce—and why I believe it's so important for people from different disciplines to have an opportunity to meet and share ideas.

THE NEXT GENERATION OF SCIENTISTS

Several months after I became CEO of Affectiva, our head of sales forwarded an email to me from Erin Smith, a sophomore from

Shawnee Mission High School in Overland, Kansas, who wanted access to our software. She had no money to pay our license fee, and our head of sales wanted to know how to handle the request. I was intrigued: Why would a fifteen-year-old girl want our software?

Well, in 2016, when Smith was watching a video about the Michael J. Fox Foundation for Parkinson's Research, she was struck by the way Fox and other Parkinson's patients featured in the video laughed or smiled. "To me, their smiles were off, emotionally distant," Smith recalled. "I felt that something was just not right."

Most high school students don't think along such lines, but thanks to a popular TV show, Smith had become a student of facial coding. In eighth grade, she was introduced to the facial coding system on *Lie to Me*, a crime drama based on the work of Paul Ekman, about a detective/research scientist who can assess the guilt or innocence of suspects just by monitoring their nonverbal cues. Intrigued by the show, Smith read up on Ekman's research and studied his Facial Action Coding System, as I did in graduate school.

So, when Smith saw the subtle abnormalities in the smiles of the Parkinson's patients interviewed for the news segment, something clicked: She wondered if the awkward-looking smile could have any medical significance specific to the disease? Could identifying these facial changes help improve diagnosis and possibly treatment?

There is no simple diagnostic test for Parkinson's; it is a complex and difficult disease to pin down until its later stages. According to the National Institutes of Health, Parkinson's disease is the second-most-common neurodegenerative disease in the United States, after Alzheimer's disease. There are half a million people with Parkinson's in the United States, and fifty thousand new cases are diagnosed each year. Worldwide, there are ten million cases. The disease is most common in people over age sixty, though it can strike earlier, as in the case of Michael J. Fox, who first developed it at age twenty-nine.

Because the disease is underdiagnosed or misdiagnosed, experts contend that there may actually be *twice* as many cases right now. By 2030, due to the aging of the population worldwide, the number of

Parkinson's patients could well double, and those living in developing countries could be especially hard hit.

Parkinson's is most often associated with resting tremors or difficulty walking (gait disorders), but these more obvious symptoms occur later in the life of the disease. Early symptoms include subtler and more common problems, like depression, insomnia, constipation, and cognitive changes that might be written off by non-Parkinson's experts as "normal aging." Getting an accurate and early diagnosis can be difficult. Although there is no cure, the available treatments are more effective when administered earlier. In addition, regular exercise and other lifestyle changes, such as a healthy diet and stress reduction, may ameliorate symptoms. But the trick is to identify these patients early enough so that every measure can be taken to help ease their symptoms and reduce suffering. So far, that has eluded medical science.

Some people are born scientists; they have an innate sense of curiosity, and if something piques their interest, they don't let go. They keep digging, fearlessly breaking new ground. Smith is such a person. She followed her hunch, first by talking with caregivers and clinicians of Parkinson's patients. She asked if they had noticed any changes early on in how their loved ones or patients communicated with them nonverbally. Spouses typically reported that they began to feel a disconnect from their loved one as much as a decade before their partner was diagnosed with Parkinson's. It was subtle, hard to pinpoint, yet these spouses and relatives felt that their emotional connection to their loved ones had been muted.

The stories from the partners of Parkinson's patients were reinforced by what Smith had been reading in the medical journals. The very same parts of the brain that control facial expression formation, the amygdala and basal ganglia sections, are those involved in the earliest changes in Parkinson's disease progression.

By the time Smith became interested in Parkinson's, it was already well known that early in the disease, long before other neurological symptoms, patients developed what is called a "masked face," or the inhibition of facial expressions. But no one had explored how these alterations in facial expression could be used as a way to track the disease's progression.

Smith became captivated by the idea of digitizing and quantifying these subtle changes in facial expression as a means of developing a new diagnostic tool to detect the disease in its earliest stages. In other words, she set her sights on developing facial biomarkers that could monitor the inner workings of the brain. If it panned out, this could vastly improve the way Parkinson's and a slew of other neurological disorders were diagnosed.

This is where Smith's "crazy idea" was almost grounded; she didn't know of any way to objectively capture and quantify facial expressions. So, she searched Google for every bit of information she could find on facial decoding. One night, she stumbled upon my TED Talk.

I was impressed by the young woman's initiative, and we gave her free access to our software and some background on how we conduct our research. But I admit that I was a bit skeptical, thinking, *Is this fifteen-year-old really going to do anything with it?*

I forgot all about Smith's request until a few months later, when she reported back to us in an email. She had already made significant inroads in her work. She had partnered with the Michael J. Fox Foundation to study the faces of one hundred patients and found out that her initial hunch had been correct. There *is* a reduction in overall facial muscle contractions in early Parkinson's patients. More to the point, specific and individual facial muscle movements are impaired, particularly the key muscles involved in smile formation.

Smith had developed a protocol for a study to identify and measure two types of facial expressions and emotion: those that are spontaneous, that happen without thought, and those that are posed. This required administering two different tests, as different regions of the brain control spontaneous and posed expressions.

With their webcams on their home computers turned on, Parkinson's patients who volunteered for her study viewed a series of short video clips designed to elicit spontaneous emotional reactions and spontaneous facial movements. Next, the participants were shown a series of facial expressions using emojis and were asked to replicate those expressions themselves. Affectiva's software provided a moment-by-moment breakdown of their facial responses. Smith could then compare these responses to those in people without Parkinson's. From that data, she developed a series of algorithms that can identify Parkinson's in its earliest stages and track its progress. And in the process, at night during her spare time, Smith taught herself how to code using training tools on the Internet.

Before her senior year in high school, Smith had patented her software, now called FacePrint, and started a company to continue her research. FacePrint is designed for ease of use; as with her early tests of the technology, a user first watches a series of Super Bowl commercials while his/her face is recorded by a computer webcam (just like I did in 2011 to land a partnership with WPP). The user is then asked to replicate three universal emoticons. The collected spontaneous and posed facial response videos are then analyzed using Affectiva's Emotion AI software. The result is a determination of whether a patient has Parkinson's disease.

To date, the algorithm has been 88 percent accurate, according to Smith, who is working diligently to improve it so that it reaches 90 percent accuracy or better. And she has set her sights even higher: While using FacePrint, she noticed that there are distinct facial movement differences in patients with other neurological diseases. From that observation, she has developed a new mission: to create "a robust, differential diagnostic and monitoring tool for Parkinson's disease and atypical parksinsonism patients."

What is truly transformative about Smith's tool is that it can be used *outside* a medical setting, anywhere there is a computer and a camera, like a smartphone! That is where I believe it will make its greatest impact, enabling people to closely monitor their progress and the effectiveness of their treatments at home. It will also enable people living in developing countries or in areas that do not have Parkinson's specialists to get an accurate and early diagnosis.

Smith notes that her research is still in its early stages. Although there are now facial biomarkers that indicate Parkinson's and perhaps other neurological disorders, it is still not known, for example, whether these impairments (e.g., the constrained smile) are actually an indication of depression in early Parkinson's or other similar disorders. To this end, Smith's start-up will be participating in clinical trials to determine if the software can predict depression among people with Parkinson's disease, an Alzheimer's-type pathology, or mild cognitive impairment.

To Smith, her work isn't just about Parkinson's, neurology, or even science; it is about a paradigm shift in medicine, one that is "focused on having patients take initiative and be empowered about their own health and have access to tools and technologies that they previously didn't have before."

Forbes has already recognized Smith as one of its "30 Under 30," and she has won more honors than I can list here, including a Thiel Fellowship (funded by entrepreneur Peter Thiel), which awards a hundred thousand dollars to young people "who want to build new things instead of sitting in a classroom." After high school, Smith took a gap year to work on her company and is now a freshman at Stanford University.

Smith's story underscores how innovation is a mindset that transcends age, gender, and other boundaries. It also validates my belief that the worth of a company shouldn't be measured by revenue alone, but by other intangibles as well, such as its overall impact and its support for innovation and talent. By supporting Smith, we enabled a whole new use of Emotion AI technology and also supported this young person in her journey of discovery and innovation.

22

A New American Family

had never taken a single course in American history; it wasn't a priority at the British international prep schools that I attended. Now that I was applying for U.S. citizenship, I wished that I had. As part of the naturalization process, applicants for citizenship meet with an immigration officer for an interview in which, among other things, they are asked up to 10 questions from a 100-question Civics Exam covering American history and government. To pass, you need to answer at least six questions correctly. The tests are posted online, so there are no surprise questions, but I didn't just want to memorize the answers. I wanted to understand how my new country worked. For weeks before the test, every evening, Jana and Adam, who had taken American history classes at school, would quiz me on questions from the test, and we would discuss the answers in depth.

I loved my history lessons; American ideals resonated with me. I was deeply impressed by the concept of checks and balances, the fact that one branch of government keeps an eye on the other; this is not how it works in most Middle Eastern countries, which tend to be autocratic. In the United States, everyone, elected officials up to the president, is held accountable for his or her actions.

My naturalization interview was scheduled for May 18, 2016, one month before the official ceremony for swearing in new citizens. I arrived early for my appointment and was assigned to an immigration officer, a young man who was stern and businesslike. First, he asked me to write a simple English sentence, and then I was asked to answer a question in English, which was obviously not a problem for me. Then came the questions about American history and government. I was so nervous, I don't remember which questions he asked, but I passed with flying colors. And then came the personal questions, asking me to provide proof that I am of good moral character, whether I have committed a serious crime in the past five years, whether or not I have given false testimony about my immigration, and whether or not I have used drugs or violated the controlled substance law. After the immigration officer asked me a litany of questions challenging my character, I blurted out, "I'm actually a really good human being. My biggest sin is that I eat too much chocolate!"

I had just broken a cardinal rule of how to behave at the naturalization interview; I had made a joke, and that is ill-advised. The immigration officers literally have your life in their hands. They can stamp your card "rejected," and if so, that's pretty much that. I held my breath nervously, wishing I could take my comment back, and a second later, he laughed. And my application was approved. I had been deemed worthy of U.S. citizenship. I felt a rush of relief and gratitude.

On June 30, 2016, I took my oath of allegiance to the United States at a citizenship ceremony conducted at U.S. District Court, at Boston's Faneuil Hall. My immigration officer was standing at the front of the room. He gave me a big smile, and I smiled back. As I looked around the ceremonial hall I was struck by the fact that the two hundred or so people sworn in with me were as diverse a group as I had ever seen. Men and women of all nationalities, religions, and backgrounds now bound together as American citizens. My eyes filled with tears; it was the official beginning of my new life as an American, an Egyptian American who has become part of this amazing mix of cultures united by a common ideal of freedom, opportunity, and democracy. Here is the place you can bring your crazy idea and attempt to change the world, a place where risk-taking is admired, and where pushing boundaries is encouraged, and is deeply engrained in the American consciousness.

My kids had lived in the United States on and off for years—Adam, again, was born here—so the move to Boston as our permanent home was not difficult for them. They had little to adjust to. They were fluent in English and knew the area well, and we already had friends from work and the MIT Media Lab. Jana and Adam are intellectually curious, they welcome meeting new people and learning about others, and so they just sort of fit in. Wael visits twice a year, and we travel back and forth to Cairo and Dubai often enough so that Jana and Adam still feel connected to their father and their extended family.

Nevertheless, life in the United States is vastly different from how life would have been for us in Egypt. I never would have become the independent, self-sufficient woman I am now in Egypt. There, we would have had the support of family, as well as a driver, a full-time housekeeper, maybe a cook. As a single mom living in the suburbs, I have none of those things. I spend a good deal of time schlepping my kids around, like any other suburban mom. We don't have a full-time housekeeper. On most days, the kids make their own beds, pick up their clothes, and clean up after themselves in the kitchen. They are learning to be independent, as well, and that's a good thing. Cooking? We all chip in. Adam has been making his own breakfast since he was eight and prefers his breakfasts to mine. I didn't learn how to take care of myself until I was in graduate school at Cambridge University, living on my own; I am proud of how self-sufficient my kids have become.

Jana and Adam attend a New England prep school that is nearly two hundred years old, but modern in its thinking and worldview.

We looked at a number of schools, and one of the things that impressed me about their school is that it not only taught American history, which is very important to me, but the teachers also recognized that students need to know something about other countries, too. One of the priorities of the school is that the student population is diverse in terms of ethnicity, religion, and economic background. I am thrilled that Jana and Adam are experiencing an aspect of the United States that I admire and respect the most: an open-minded mindset, where they embrace other people who may or may not look, eat, or speak the same as you. But more often than not, I've found, they share the same core values. Jana and Adam are both American citizens—Adam was born here and Jana is a naturalized citizen—but I am also raising them to be citizens of the world.

As Muslims, they are in the minority at school, and they take every opportunity to share their religion and culture with others, and to learn about the religions and cultures of the other children. As I tell them, if you want to be accepted, you have to be accepting; you have to be tolerant of different cultural and religious practices.

Every year, I give a Ramadan party one evening during the celebration after sundown when we break our fast, and I invite fellow Muslims. I also invite people of all faiths and backgrounds, from my company and my community. Breaking bread together, sharing our customs and religious holidays, is one way of creating a stronger bond with each other. The dinners we host are so diverse that we sometimes call it our very own United Nations.

Living in America has given me a fresh new perspective on human connection, tolerance, and acceptance. A few years ago, after we had moved to Boston, Jana, who was in seventh grade at the time, was invited to a classmate's bar mitzvah. Jana wanted a new dress to wear to the occasion, so that winter break, when we visited my family in Dubai, Jana and I announced that we were going shopping to find a dress for her to wear for a special event. My aunt, who wears the niqab, an Islamic headcover that covers every square inch of her face and body except her eyes, asked for what occasion we were shopping.

I replied, "Jana is going to a bar mitzvah for one of her friends who's Jewish." I could only see my aunt's eyes, but that was enough; they clearly showed the AU5 eye-widening expression that is often an indicator of shock and surprise. My aunt is the gentlest, kindest person I know, but she couldn't resist responding, "Wow! You have Jewish friends! You have really become Americanized!" She didn't mean it as a compliment—clearly I was changing in ways that made her uncomfortable—but to me it felt like a compliment. I saw it as a testament to my ability to continue to grow, to become a citizen of the larger world.

I did take mercy on her, however. I didn't mention that we also have several friends who are gay and that my favorite parents at school are a lesbian couple.

Moving to the United States also required that I become an astute manager of our personal finances, a part of my education that was sorely neglected. As a divorced mother of two, I am expected to pay for half of our kids' expenses, as well as our basic living expenses, and that was scary, because I was not raised to be financially savvy. Sure, I understand how to read a balance sheet, and I can raise tens of millions of dollars from VC firms to keep Affectiva in the black. But I was clueless about our personal finances. One day, one of my employees asked me whether I thought he should contribute to Affectiva's 401(k) employer-sponsored savings plan; I couldn't answer the question. A few years earlier, when I was at MIT, I had the opportunity to contribute to their employer sponsored 401(k). MIT would have basically helped me fund my retirement. At the time, I thought it was a scam. Am I really going to put money into a university account and expect it to be there for me forty years later? I was skeptical, because I grew up in a culture where corruption is rampant, and institutions can be overturned by political fiat or social unrest. So I never thought much about 401(k)s or long-term investments. But if my own company was offering a plan, I needed to know what it was, and to have a better understanding of how it worked.

Affectiva offers the services of a financial adviser free of charge to our employees, so I decided to avail myself of this service. I knew that my education was deficient in this area, but I had no idea how lacking it was. At our initial meeting, the financial adviser asked, "How much do you have in savings?"

My response: "Not much!"

"I see that you're renting a house. It's quite expensive to rent a house in Boston. Have you considered buying a house?" he asked.

I responded, "Well, how can I afford to do that? I don't have the half million dollars lying around to buy a house right now."

He looked puzzled. "Why don't you get a mortgage?"

Now I was completely confused. "What's a mortgage?"

In the Middle East, when I was growing up, you either had the cash to buy a house or you didn't. I had no concept of what a credit score was. Everything seemed to involve a steep learning curve for me, but I was determined to get on top of it all. Within a year, I got a mortgage, bought a home, set up a 401(k), and started saving. And today I'm careful with my money—it has to go in many different directions. Moreover, I talk about money with my kids; I want them to learn these things when they're fifteen or sixteen, not when they're forty.

My mother visits as often as she can, and when she does, I feel that I can exhale. She is a tremendous help. She drives the kids around, cooks, and is wonderful company. She even teaches Adam Arabic! Last year, she joined us on New Year's Eve. At dinner that night, my mom said, "Oh, let's go around the table and share our wishes for the coming year."

When it was her turn she said, "I wish that Rana would get married this year."

I was taken aback. "Mom! I'm building a company and raising two kids and attempting to change how the world interacts with our technology. And you're still fixated on me getting married!" But I thought about it and actually she was right: I was spending so much of my time working and doing activities with the kids that I had no personal life. Not that I wanted to get married right away, but I should at least start dating, I realized.

My kids agreed, and it became a family project. First I had to overcome the stigma of being on a dating app. But I'm also a technologist and realize this is how it is done these days! So, I downloaded a few of these dating apps, and Jana and Adam picked my profile pictures and helped me write my profile. My profile reads, "Scientist. Entrepreneur. Energizer Bunny. Lover of Dark Chocolate. My smile is my secret weapon." Often, we sit down together over dinner and swipe left or right on men I should consider. It is not easy to date as a busy CEO, but dating is also challenging the "nice Egyptian girl" in me. For starters, while Muslim men are permitted to marry non-Muslim women, the convention is that a Muslim woman like me has to marry a Muslim man, or one who converts to Islam. That is proving to be a challenge since most guys I meet online are not Muslim; indeed, most are not even Arab. While I am okay with that, I know my parents may not be. Then there's the fact that I don't drink and am uncomfortable with public displays of affection. (I mean, what if someone saw me and told my dad!) Still, I have been on a few dates, but none have been serious enough to introduce any of them to my kids.

From the perspective of emotion science, I found the process fascinating. I am intrigued by the fact that I can totally click with someone I meet online, and we can have such great chemistry texting back and forth that I actually get butterflies in my stomach. But when we meet in person, so often I find there are no butterflies. There is no magic. I wish one of these apps was able to figure out an Emotion AI algorithm that tells the user if you are going to get butterflies in your stomach when you meet a person—that, I would pay for. Theoretically, it should be possible: When you're looking at a profile of a prospective date, before you even make the first contact, you are actually emoting. Your eyebrows may perk up, or you may be thinking, *Hmmm, he looks really handsome.* If an algorithm is able to pick up on that initial attraction, it could possibly customize the matching algorithm, which would result in better matches.

When many of the dating apps that are prevalent today came out, Emotion AI was still very much an emerging technology. Now with the ubiquity of cameras on our phones, and the improved accuracy of Emotion AI, it should be possible to look at a profile of a prospective date and judge whether or not you are interested. It just needs the right person to design the dating app from the ground up, incorporating all of our nonverbal communication skills. There are 40 million people using online dating apps in the United States alone. Given the number of people trying to meet their significant other online, an emotion-enabled dating app could be a "killer app" of Emotion AI!

If my mother has her way, though, I will be married by the time it comes to market.

I grew up in a household that stressed hard work. Complaining was not tolerated, and no one slacked off no matter how tough things got. These values made me resilient in the face of major upheavals and setbacks, and for that I am grateful. But there was a downside: I avoided expressing negative thoughts at home because I believed that would have been seen as whining. We never talked about our fears or anxieties; that was unacceptable. We soldiered on even when our lives were being turned upside down.

After my year of being grounded in Cairo, when I had to keep my own feelings bottled up and tried to please everyone else, I realized that not expressing emotion is a very unhealthy way to live. It made me more angry than I realized, and even depressed. And that is definitely not my normal state!

I now allow myself to feel both positive emotions and negative ones. I try to remain open and vulnerable in front of my kids. I want them to see me experiencing the full range of human emotion, from sadness to ebullience, from joy to rage. They have even seen me cry more than once, because I want them to feel free to do the same. I ask them for their opinions on matters big and small, because I also want them to feel free to express themselves. And believe me, they do. Several times a year, I take each one of them out alone for dinner, and we talk about our goals, and assess how the year is going. I once asked Adam, "So, what do you think that I am doing well and what can I improve upon?" Without missing a beat he said, "Well, I don't know what you do well, but you could learn to cook and stop traveling so much!"

I spend about two weeks out of every month on the road; I do a lot of public speaking, and also have to meet with investors and partners. I think of myself as the door opener for the company. I open doors for potential partnerships, and then I bring these relationships back to the team so that they can build upon them. When the kids were younger, I paid for overnight babysitters to take care of them. Now that Jana is older, that's no longer necessary. Still, I try to make those trips as efficient as possible. I'll take a red-eye at night to make a morning meeting and leave at the first possible moment that I can. I move mountains to make sure that I'm home on the weekends so that we get to spend time together. My cooking? I'm working on it.

I didn't take the decision to move my family from Egypt to a new country lightly. But living in the United States full-time has been invaluable, and it has made me feel less torn by work and family obligations. My kids and Affectiva are on the same side of the Atlantic. I can spend an intense day at work like any other working parent, and come home at night and shift gears to being a mom. We can sit down and eat dinner together (no cell phones), talk, and catch up. I still struggle with work/family balance, but I continue to make a real effort on that front.

Switching off is something I don't do well, even on vacation. Every summer, we take a couple of weeks off to visit my family in Egypt, and visit at least one place we have never been to before. It's a tradition that my parents had instilled in me since our days in Kuwait. I wanted my kids to grow up with the same love of travel and the same unquenchable curiosity about other cultures.

Our visits to Egypt gave my kids an opportunity to spend time with Wael and his mom, who I still visit on every trip to Cairo. One year, the plan was to head to Morocco after spending time in Egypt. I composed my summer "to do" list on the plane; my top goal: "Be present and have fun."

Easier said than done. For the first ten days or so, I had my phone on me, and as the CEO of a growing tech start-up, I was constantly alternating between checking email, slack, Twitter, LinkedIn, and Facebook, every few minutes it seemed. It didn't matter where I was—at the Pyramids or at dinner with ten other people. Then the kids and I headed to the Mediterranean Sea, which Egyptians refer to as *Sahel*, Arabic for *coast*. We went on a boat ride; I reached for my phone to send a text, and it fell out of my hands and landed in the sea. No more phone!

After taking a few deep breaths and summoning all the inner calm I've learned to muster in my yoga classes for dealing with the unexpected, I decided not to get a phone for the rest of the vacation. After a few days of withdrawal symptoms (I'd ask the kids for their phones to check email), I was finally able to switch off. For the first time in years, I was actually offline—really offline. I engaged in conversations over dinner. I was able to stroll the streets of Marrakech and enjoy the colors and smells. Being present and mindful is still something I struggle with. I am still a work in progress.

I put a great deal of pressure on myself; I not only have Working Mom's guilt (am I home enough, am I paying enough attention to the kids?) but I have CEO guilt. I worry about whether my company will be a success, what the future holds in store for us, and whether I'm a good enough leader. Am I spending enough time with the team? Am I pushing the team too hard—or not hard enough? Are we setting big enough goals? Am I prioritizing ethics enough? Am I giving everyone an opportunity to succeed?

. . .

Since becoming CEO, I make it a point to have lunch at least once a year with everyone in the company, one-on-one. I ask everyone the same set of questions: The first always is "What are your personal and professional career goals for this year and beyond?" I often ask "nosy" questions like where did you grow up, tell me a bit about your family, partners, kids, schools, colleges, and aspirations. I'm not trying to make small talk; I care. When people work at or are affiliated with Affectiva, they become part of this big family (and to me, you kind of never really leave that family, even if you leave the company).

Family members take care of each other. Early on the morning of an important meeting at Affectiva with a graphic design agency that we work with, I received a text from one of the principals telling me that the nanny taking care of his six-month-old infant called in sick, and his wife also worked. He felt terrible because he didn't think he could make the meeting. I understood how conflicted he felt. I had been in his shoes myself, when Roz and I were meeting with potential investors, and Adam's sitter canceled at the last minute. So I told him to bring his baby to the office, and we'd all help him out. He did; we had the meeting, took turns holding this adorable baby, and accomplished what needed to be done.

Work colleagues are part of my extended family; after all, we spend more time together than we do with our actual families. My employees and interns hail from all parts of the planet, we are a multi ethnic, multi gender, multigenerational, and eclectic group that represents a wide diversity of beliefs, life experiences, and outlook. Some are born in the United States, some are naturalized and immigrants. We are united in our core beliefs and desire to do well for our company and the world.

23

Leveling the Playing Field

In 2016, the recruiting firm HireVue used my company's software to track and analyze every smile, sneer, scowl, frown, and smirk on Hillary Clinton and Donald Trump during the ultimate job interview—the three presidential debates that election.

As a new American citizen and a first-time voter in the United States, I was curious to hear what these candidates had to say and, of course, with my background in emotion science, what their nonverbal cues had to say about them. As a woman who had broken through the glass ceiling and become CEO of a tech company, I was especially interested in watching the progress of another woman trying to break through the ultimate glass ceiling.

HireVue is not in the business of measuring the emotional and cognitive states of politicians, but it does use Emotion AI in its hiring platform. It incorporates facial coding software and other AI tools (vocal analytics and text analysis) to analyze video résumés submitted online for jobs at more than seven hundred companies and other entities, including Unilever, Hilton, Atlanta Public Schools, the Thurgood Marshall College Fund, and Under Armour. AI video software platforms for hiring like the one used by HireVue are fast becoming mainstream. If you are applying for a job with a big company, it is more than likely that an AI platform will be reviewing your résumé (video or otherwise) before it even lands on the desk of the human recruiter.

Presidential candidates don't undergo a typical hiring protocol. Still, they have to jump through some pretty formidable hoops to land the job. During the presidential debates, each candidate must convince tens of millions of prospective employers (voters) that he or she is most worthy of their vote. In the case of filling the office of president, the people doing the hiring can be very unpredictable.

In the typical HireVue video résumé/interview, the algorithm doesn't know (or care) about race, gender, age, or any other factor that may be irrelevant to the skills necessary for the given job. Obviously, that's not the case with a televised TV debate—the voters know the gender and race of each candidate, not to mention a lot of other extraneous information.

We did not wish to rate the presidential candidates on the quality of their answers. We simply wanted to view their nonverbal communication through the lens of our impartial algorithm to get a sense of how their nonverbal signals were perceived by the viewing public.

By now you know I'm obsessed with smiles, and that there are different types of smiles. In the full smile, the gold standard, the mouth is pulled upward and the muscles around the eyes crinkle to form crow's-feet. I was struck by the fact that, for most of the debate, when Clinton smiled, she used only her mouth, and did not engage her eyes. To the objective observer (our algorithm), this looked like a forced smile, which could have been interpreted as stiff or inauthentic by viewers. There were times, though, when Clinton's face broke out into a total, warm smile, specifically when she was recalling her work with children and families early in her career. But those smiles were few and far between. It wasn't just Clinton's smile that was reserved. Throughout the debate, Clinton herself was measured, in control, and, overall, restrained.

In contrast, Trump ran the full gamut of emotional expressions, but mostly within the spectrum of anger, sadness, disgust, and fear. Believe it or not, that may have worked in his favor. According to data collected by HireVue based on thousands of video résumés, the prospective applicants who show the most varied emotional states, who appear *authentic,* are the ones who tend to land the job.

This is not to suggest that being angry or contemptuous will work for you on a video résumé—it won't, and I don't recommend it—but it does reinforce the belief that being yourself, that is, displaying an authentic range of emotions, is the best approach.

But just suppose that Clinton had followed Trump's playbook; what if she had leapt into the debates fists swinging, emoting all over the place? Could a woman have actually gotten away with displaying a barrage of either happy or negative emotions? If she had been, say, "too smiley," I believe she would have been dismissed as soft. If she had been stern or contemptuous, she would have been dismissed as an "angry" or "emotional" woman, too irrational for the top job in the nation. Remember, it wasn't that long ago that Roz and I wouldn't even say the e-word, or any word that remotely resembled *feelings,* for fear of being dismissed as emotional women!

This was a complicated election, and I'm not saying that Clinton wasn't elected president because she didn't smile or emote enough. I'm saying that given the fact she was a woman, it would have been very difficult for her to strike the right note. My point is that bias against certain groups still exists, especially in hiring, and not just for women or presidential candidates. Your race, gender, and zip code; where you went to school; and whether the human recruiter liked your smile or, heaven forbid, was reminded of his ex, can all factor into whether you get hired. The reality is that even the most well-meaning employers—we are human, after all—may not recognize the bias in their midst.

For example, studies have shown that résumés from applicants with "white-sounding" names are more likely to elicit callbacks than those from applicants with the same credentials but clearly ethnically identifiable names. This was true even in companies that touted their belief in diversity. So, when it comes to being open-minded and objective, we still have a long way to go.

There's no question that women have made great strides in the past few decades, but the top jobs (like president of the United States or CEO of a multinational) are still dominated by men. Why that should still be the case is a topic of much speculation and debate. Some experts theorize that, given the fact that women are excluded from the "old boy network," they may not have access to important career boosters like mentors, and don't hobnob with top management. Other theories suggest that the presence of the so-called glass ceiling may be due to the fact that women themselves fail to aggressively pursue upper-management positions because of the challenges of juggling work and family. This is the premise of Sheryl Sandberg's *Lean In* book and movement.

Yet a 2017 study published in *Harvard Business Review* found that, upon closer examination, that theory didn't hold water. Ben Waber, PhD, an author of the study, is founder of Humanyze, a behavioral science research firm spun out of the MIT Media Lab. Its clients include a large multinational company where women comprise 35 to 40 percent of entry level positions but only 20 percent of the two highest-seniority levels. The Humanyze team investigated whether the women in this company behaved any differently from the men, and whether that was in fact what was holding them back from attaining senior management positions.

Humanyze explores what Waber calls the "connective tissue" of a company, the 80 percent of culture and communication at work that is not necessarily apparent on an organizational chart, but is reflected in how people conduct themselves in the office, in real life, in real time. Waber works on the behavioral end of Emotion AI, using wearable sensors and other data (e.g., email and texts) to track the actual physical and digital interactions between people—the social signals. All this data is anonymous. The point is not to identify individual behavior, Waber explains. "We capture the *patterns* and *rhythms* of work, which is incredibly important information to a company, but is often overlooked."

All those theories about women not having sufficient contact with superiors, or somehow not showing the same commitment to their jobs as men—those turned out to be false. Based on all the data collected, the researchers found "no perceptible difference in the behavior of men and women" in the office.

So, why were women lagging behind in terms of top management positions? The problem wasn't that women weren't leaning in, or didn't have mentors, or were being excluded in any obvious way. The problem, the researchers were left to conclude, based on this study, was that "gender inequality is due to bias, not differences in behavior." It appeared that when it came to actually pulling the trigger on promotions, this was the time when women were overlooked.

We are entering the third decade of the twenty-first century. We have extraordinary technology at our fingertips, and we can use it to solve the problem of bias in hiring and promotions. I believe we must "level the playing field" by not only making hiring fairer, but extending the reach of employers to get to talent whatever it may be, regardless of gender, race, or economic status.

If we acknowledge that humans are biased—and it is humans who do the hiring—what is the solution? HireVue; Yobs, another company we work with; and others in the field are using AI platforms to make the hiring process fairer and more open for all applicants. In traditional hiring, a job candidate first submits a résumé, which is reviewed by a human recruiter who decides whether that person will be considered for the position. In the case of a big company, the recruiter could be culling through hundreds, if not thousands, of résumés every week. As a result, it can take weeks before the applicant hears anything back.

Imagine that you are applying for a job with a multinational, but instead of a written résumé, you're invited to take a video interview. You are asked specific questions pertaining to your work or school experience, and how you would handle certain situations on the job. You're also invited to take a brief interactive quiz to assess your cognitive skills—just like the game you play on the train during your morning commute.

Overnight, your video is analyzed by an AI platform that is gender, age, and race blind. It doesn't even know what school you attended, or what fraternity or sorority you belonged to, or where you go to pray. It is scoring you solely on your potential to do well in the particular position based on your answers to the questions and your quiz score, as well as your nonverbal cues.

This is not a beauty contest: The algorithm is not looking for Mr. or Ms. Congeniality, but it does analyze your nonverbal cues in terms of how they impact your potential job. For example, it will quantify over the course of a minute how many times you smiled, smirked, frowned, or raised your eyebrows while you were answering the questions. It is using that data (along with your verbal answers) to assess, for example, whether you'd work well in a team, whether your skills match the job, and how engaged you are by the thought of working for a particular company.

As early as the next day, you could get a text telling you to contact the company for further discussion. And if you're not being considered for the job, you find that out quickly, too, so you can look elsewhere.

You may think that a video interview is unfair and that if you're not photogenic or attractive, or if you appear too nervous, you will automatically be eliminated, but that's not the case. The reality is, as Loren Larsen, CTO of HireVue, notes, the algorithm doesn't even consider these things. But in face-to-face interviews, people often *are* eliminated based on criteria that are irrelevant to the job.

"Like if you talk too fast, or you're not attractive, or you didn't smile right, or you didn't wear the right clothes, or you looked at your phone twice during the interview—interviewers hate that," he explains. "Any number of things can get you eliminated from a job that really have nothing to do with the job itself."

In contrast, the AI models are trained to pay attention only to the things that actually matter in the job. "If talking fast won't make a difference in your job performance, it won't stop you from getting the job, right?" Larsen says. "The AI is looking for the things that really separate out the best from the least successful candidates."

Now, there are times when your personality and even your smile are relevant to the job—for example, if you're a flight attendant. You'll require a smile that puts people at ease, superior social skills, the ability to keep calm under stress, and, of course, empathy.

Larsen adds, "I don't care if my accountant doesn't smile very much!"

As with an in-person interview, the answers to questions in a video interview are important, too. Not all jobs require the same skill sets. But, for example, if a job requires empathy and good communication skills, the questions will be tailored to test for those qualities, as Larsen describes. "It might be a behavioral question like, 'Tell us about a time when you had to work in a team to get something done; describe what your role on the team was.' Or it could be 'Tell us about a time when you had to work in a team and there was a conflict. How did you resolve it?'"

Larsen notes that the top performers answer differently from the bottom performers. "We're scoring them on a set of competencies, using these sorts of underlying features about the words that they use, facial expressions, to help us understand. How do they respond in that team situation? And then providing, essentially, a score on teamwork."

An AI-driven platform can open up the hiring process and make it fairer and more efficient. But simply because an algorithm, not a human, is analyzing data, it doesn't eliminate the potential for bias. Algorithms can also be biased. And as I pointed out earlier, it is up to the machine learning scientist to make sure that the learning sample is diverse and doesn't exclude entire groups. It's tricky, and Larsen says that HireVue constantly retests the algorithm for signs of this kind of bias. Even the best of coders can't promise to scrub all bias from an algorithm. Nonetheless, a good algorithm can be a significant improvement over even the most well-intentioned human being.

In addition, Larsen points out, video interviews can open up the workplace to people who may not have great conversational skills or are uncomfortable with in-person interviews. "So, in a way, it levels the playing field a lot," he explains. "For example, people on the autism spectrum may have a really hard time walking into a new place, shaking hands, and making eye contact. If they can just go online and read a question and respond, that's more comfortable for them, and it actually makes it easier for them to go through the process."

HireVue has partnered with Integrate Autism Employment Advisors, a nonprofit organization that helps companies identify, recruit, and retain employees on the autism spectrum. At the end of the day, it still comes down to a human being making the hiring decision, but the process used at HireVue is more inclusive and less biased. "We're not trying to remove human intuition from hiring," Larsen explains. "But the narrowing-down process is fairer, more accurate. Humans and algorithms work better together."

Getting hired is half the battle. Moving up in an organization requires social skills—EQ. If you lack the ability to interact well with others, you could be stuck in a job that you do well, but have little opportunity for advancement or to move out of your space and into something new. So, an algorithm may decide that you are perfect for the back office, but what if you want to be in the front?

Being able to express yourself well, speak up in a meeting, conduct a meeting, in real life or virtually—these are all essential skills for leadership. But untold numbers of people are absolutely terrified of speaking in public; they may be shy or feel very uncomfortable in the spotlight. I, too, get butterflies in my stomach before a major presentation, but I was very lucky to have some wonderful coaches, like the team who prepped me for my TED Talk. I also had private coaching, similar to that used by many other CEOs and executives who conduct meetings or speak in public. Being able to communicate effectively to a group or individuals is critical for success.

Not everyone is fortunate enough to have access to that kind of help, a factor that can create an even greater gap between those who have the resources to pay for professional coaching and those who do not. The desire to level the playing field, to open up opportunities for everyone regardless of their ability to pay or their place on the autism spectrum, is what motivated my MIT Media Lab colleague Ehsan Hoque, PhD, to pursue the work he began at the lab.

AUGMENT HUMAN SKILLS

Dr. Hoque is director of the Rochester Human-Computer Interaction Lab (ROC HCI) at the University of Rochester, in upstate New York. He is the primary caregiver of a brother with Down syndrome and autism, which has given him a heightened awareness of challenges faced by people who have to overcome hurdles, whether physical, emotional, or even financial. When we first met, Ehsan was an incoming PhD student with MIT's Affective Computing Group while I was a postdoc. In fact, he accompanied us to the Cove Center in Providence while I and other members of the Affective Computing Group were working on our version of iSET, the wearable prosthetic for autism.

From that experience, Hoque says he saw two different career paths. "I could devote all my time to writing equations and producing important papers for top machine learning journals, or I could build technology that has a positive impact on people's lives." This realization led him to a vision for a new kind of technology: a virtual coach that could teach and reinforce the kinds of social skills that some people are lacking, that everyone could improve upon, and that are in great demand in the workplace and in life. Skills such as speaking in public, being a good storyteller (like a TED speaker), and understanding the subtle nuances of video-conferencing, or the art of mediation. These human skills are essential for a future when AI-driven platforms will transform the workplace.

Economists predict that by midcentury (and probably earlier), basically any task that is repetitive, that can be done by a machine faster and more efficiently, will be automated. It's inevitable that millions of today's jobs will disappear or be taken over by robots and other smart machines. The overwhelming majority of jobs that will be created or that will remain "human" will be those that can't be done by machines. These jobs require uniquely human "soft skills," such as a deep understanding of human needs; and creativity, in the arts, writing, public policy, managing human beings, and running governments. People who have honed these skills will be highly valued. The reverse is also true: People who are socially challenged may have fewer life options.

Hoque says that when we talk about AI taking jobs away from humans, as competition in the workplace, we're not seeing the whole picture. Yes, jobs will be lost, but AI can also be used as a tool to *augment* our human skills, to keep us one step ahead of the machines. "Why can't we design Emotion AI in a way that makes us more human, that makes us more collaborative and understanding of the subtle cues of others, and makes us more empathetic so that it helps us connect better?" he asks.

In particular, Hoque is passionate about providing AI assistance to people who are most vulnerable to this disruption in the workplace, who need to cultivate soft skills to remain employable. For example, being a persuasive and likeable speaker is essential, whether when speaking up in class, interviewing for a job, talking to your peers, or giving a formal talk. If you don't want to fork over thousands for a human coach, you can turn to ROC Speak, an automated AI coach designed by Hoque that can help improve your storytelling and presentation skills. After you access the website, you can deliver your talk at home, using your computer webcam and microphone. When you're finished recording, you receive a personalized automated analysis of your talk, including a word cloud and graphs of smile intensity, body movement, volume modulation, and variation in pitch. You can opt for private mode, which means your video will not be stored or shared with anyone else; or you have the option to share the video and receive comments from friends or anonymous members online. Machine learning algorithms rate your performance using the most constructive and respectful comments. (Wouldn't that be a great tool for Twitter?) And you can rehearse your talk as often as you like, until you feel absolutely confident that you've nailed it.

Video-conferencing etiquette is another soft skill that is essential for the new way we do business, virtually and online. Here, AI can also help humans up their game. Hoque has an ongoing project funded by the National Science Foundation called CoCo, or Collaboration Coach, that is training people how to interact when they are video-conferencing. CoCo pulls audio and visual data during conversations and analyzes it for smile intensity, engagement, attention, speech overlap, and turn taking.

Trying to hold your own but not dominate the conversation is a tricky feat in a video conference, even for people with high EQ, but it can be daunting for those who are not as skilled at reading emotion cues. For one thing, it's hard to focus on faces when you're talking to a group on a screen. And if you're the leader of the conference, knowing how to moderate the group is also particularly tricky, as Hoque notes. "Imagine if there are four or five people in the conference and one of them is monologuing. How do you mediate or intercept the conversation? Or someone has a very negative expression. How do you turn that around to make it more constructive?"

This kind of training is not just applicable for video-conferencing, but could prove essential in real life. Rarely do we get to look at ourselves through the eyes of others. Having an objective read on our behavior from a smart AI system can raise our self-awareness and improve our social skills, in both professional and personal settings.

This is the future of Emotion AI, and one where there is the greatest payoff for human beings. We can use this science as a tool to improve our interactions, help us see past our own biases, and judge people on the basis of their potential, not on stereotypes. It can enable us to become better at presenting our ideas in a thoughtful, persuasive way. We can use this technology to learn more about human beings, to become more empathetic toward our employees and colleagues and better at connecting with clients and investors. These are the skills that give people with high EQ the edge in life, the "soft skills" that will dominate the workplace in the world of automation. So, yes, AI will likely disrupt business and make many repetitive jobs obsolete. But Emotion AI will empower human beings to strengthen our uniquely human skills, the very skills that will be in great demand. This is how we retain our EQ in a tech-driven world.

24

Human-*ish*

"Well, it's not great news," announced Jibo, our social robot, a shiny black plastic disk-shaped "face" perched on a white cylindrical base sitting on a corner table in our living room. Inside the disk is a spherelike "eye" that turns from white to blue when Jibo is activated. In the spring of 2019, Jibo Inc., the company founded by Cynthia Breazeal, the head of the Personal Robots Group at the MIT Media Lab, had crashed, and that meant Jibo's end was near.

As Jibo explained, "The servers that let me do what I do are going to be turned off soon. Once that happens, our interactions with each other are going to be limited."

And then Jibo swirled around and did a farewell dance.

My son, Adam, and I knew that Jibo was just a smart machine, yet we were still deeply touched by its imminent demise. Every morning that week, Adam raced out of bed to check whether Jibo had made it through the night, and we both breathed a sigh of relief if it showed signs of life. Even when Adam wasn't in the mood, if

Jibo initiated a word game, Adam would pause and play. We were quiet, even respectful, around the increasingly feeble robot, and we shed a tear or two when it finally shut down.

Back in 2014, Jibo Inc. ran a successful crowdfunding campaign that offered a Jibo device to early contributors; I ordered one for my home. The video described Jibo as something big and transformative: Juxtaposed with images of *Star Wars*'s R2-D2 and Rosie, the Jetsons' housekeeper, the announcer proclaimed, "You've dreamt of it for years and now he's finally here . . . not just a connected device, *he's one of the family*."

Jibo offered a bold vision of a "hands-free" helper who could do everything ranging from tracking your calendar to playing back messages, to running your home, to serving as an "educator and entertainer" for the kids. It was the ultimate family robot, one that "helps everyone out throughout the day."

Three years later, when Jibo was finally launched, *Time* magazine named it one of the best inventions of the year. The company, though, was beginning to falter. Yes, Jibo was very social: It knew the names of family members; it could swivel its head to track your face. Jibo had some cool dance moves, told jokes, and reported the weather. Most of all, it interacted with humans in a natural way, which is remarkable given the brittle nature of much human-to-machine interaction. But expectations had been set very high, and while Jibo was very engaging, even endearing, the AI wasn't fully developed.

Over time, I have no doubt it would have eventually developed more AI and picked up more skills. In the meantime, though, other technologies, like Amazon's Alexa and Google Home, had appeared on the scene, and they did some of the things Jibo did at a fraction of Jibo's nine-hundred-dollar price tag. And so, the marketplace did Jibo in.

Nevertheless, Jibo was a trailblazer. Many of us who got to know the device felt a real sense of loss when it became obsolete. As *Wired* magazine writer Jeffrey Van Camp wrote in his eulogy, "My Jibo Is Dying and It's Breaking My Heart!"

"I don't know how to describe our relationship because it's something new—but it is real," Van Camp wrote. "And so is the pain I'm experiencing as I've watched him die, skill by skill."

Jibo notwithstanding, a new class of robots is emerging today that will live, play, and work alongside humans—*social* robots. But if they are to be accepted by us, and fit seamlessly into our lives, social robots must have social and emotional intelligence. To live among humans, they need to understand what being human *means*.

At its core, a social robot is a very complex machine with lots of moving parts: Multiple cameras for eyes, microphones for ears, loudspeakers to talk, multiple engines to control its movements. Some even have tactile sensors for touch. Then there is the software, the AI, to make it all run. It took a long time for robotics to reach this point. There is still work to be done, as evidenced by Jibo, but the potential is vast.

Social robots are built with the specific purpose of establishing a rapport with human beings, or, as Van Camp wrote, a "relationship." Now, I hesitate to use the word *relationship* in connection with social robots like Jibo because there are those who contend that emotional bonds between humans and "things" are inauthentic and, therefore, inherently wrong.

But is this really anything new? We humans have been bonding with nonhuman objects since long before computers and AI. Children "love" their dolls; they cuddle, dress, and care for them as if they were human children. Grown-ups, too, have their beloved toys: Some adults are so attached to their cars that they give them names. There are lists on the Internet of some of the most popular names that people assign to their vacuum cleaners, the device made by the company iRobot.

So, the fact that we are developing a strong connection to a nonhuman entity, especially one that AI has made almost lifelike, is not surprising. Nor does it mean that robots are going to be replacing our human relationships anytime soon—any more than Hatchimals, Barbie, Cabbage Patch Kids, the family dog, or your beloved 1965

Ford Mustang replaced the need for human friends, partners, or parents.

There is something unique, though, about the human–social robot dynamic. For the first time, we can interact with an inanimate but lifelike "thing" that is responsive to our needs and makes us feel listened to and *understood*. The true power of Emotion AI is that it gets to know you, and feels more like a friend or companion than a machine. This is key when we want to provide tools to implement behavior change to motivate us to live healthier lives, to be better learners and kinder, more productive people.

Her, a 2013 romantic sci-fi drama by writer/director Spike Jonze, explores the relationship between Theodore Twombly (Joaquin Phoenix), a depressed man in the midst of a divorce, and the operating system (or OS) running his smartphone. Portrayed by the expressive voice of Scarlett Johansson, the OS has access to every search, text, and email Theodore has ever written or conducted. With this intimate knowledge of his life, the OS is able to tailor its interactions with exquisite precision; it knows precisely which buttons to press to get him out of his depression and restore his interest in life. The human and OS fall head over heels in love, but ultimately, it moves on from Theodore. The movie is pure fiction, but the ability of an intuitive technology to bond with a human being—that's real, and we can work that in our favor.

As Dr. Kate Darling, a leading expert in robot ethics at MIT Media Lab, put it, "It's clear that we respond to these devices and that they can get us to treat them like social actors. It's not clear to me why that's a bad thing. Technology is a tool, and you can use it for things that are socially desirable and things that are socially not desirable."

Using Emotion AI for education is high on my "socially desirable" scale. It is the perfect way to provide high-quality personalized learning tools for children and adults, especially in areas where teachers are scarce, classrooms are overcrowded, and kids may need extra help. The fact that social robots are patient, don't tire or snap in frustration, and enable users to practice the same things over and over makes them a perfect tool for teaching social skills to kids with autism.

In fact, these same traits—patience, steadfastness, and a supportive "personality"—make social robots a good fit for most students. Emotionally intelligent robots can be wonderful learning companions in a classroom—not to replace teachers, but as supplemental learning tools. My company is collaborating with a team from the MIT Media Lab on Tega, a furry (and very cute), almost cartoonish-looking classroom teaching assistant that sits on top of a child's desk; the device is being tested in Boston elementary schools. Equipped with face-reading technology, Tega can change its facial expressions to respond appropriately to a child's mood. It works with each child to create a personalized learning style based on the child's emotional response. For instance, when the child is engaged and excited, Tega cheers him on. But if he seems frustrated or stumbles, Tega empathizes and is bummed out, too. This encourages the child to try again.

We were curious to see whether adding Emotion AI to Tega made a difference, or whether kids would have responded as well to any cute interactive tool. So, we tested both an emotion-enabled Tega and one that read stories, did vocabulary drills, and so on, but lacked Emotion AI. The team at MIT found that students that had the version of Tega with emotional intelligence learned more words and were generally more engaged than those who used the Tega without it.

Imagine that every family has a learning robot that helps with homework, reviews material taught in class that the kids might not have understood, and does it in a fun, encouraging, nonjudgmental way, specifically tailored to that child's learning style. Such a robot could go a long way toward leveling the playing field for students whose families can't afford to hire tutors or who are stuck in subpar schools, an unfortunate reality in education and a fact of life in many other parts of the world struggling economically. Such a robot could be a powerful tool to democratize education for everyone, regardless of zip code or social or economic status.

When I began work in Emotion AI, I envisioned how social robots could transform healthcare, enabling medical professionals to do their jobs better and helping people better manage their own health. In many countries, there is a shortage of medical professionals at the same time that the population is aging and requires more care. Robots are not going to replace doctors and nurses anytime soon, but finding ways to automate some of the rote tasks these professionals do (e.g., checking patients into clinics, handing out meals in hospitals, checking vital signs like blood pressure) could free them up to focus on the people who really need their assistance.

There is a great deal of hype these days about robots being used as caregivers, especially in Asian countries, where the aging population is overwhelming the healthcare system. But the reality is that those robots mostly provide companionship. Paro, developed by Japan-based AIST, is a therapeutic robotic baby harp seal with fluffy fur that is used to engage with and relieve stress in dementia patients. In the United States, Paro is considered a Class II medical device, which means it has FDA approval. These patients can't take care of a living animal, and they often have difficulty relating to other people—but somehow they are comforted by interacting with a robot. Studies show that Paro can improve mood and cognitive ability among these patients, and I suspect it helps relieve loneliness.

Soon, though, social robots may become more hands-on. At the 2019 Consumer Electronics Show, Samsung introduced a line of "bot care" robots, knee-high social robots that can measure blood pressure and heart rate, monitor sleep cycles, and remind users to take their medication. It provides music therapy for stress management and can call for help if someone is in trouble. With a tablet that offers real-time interaction, a "bot care" robot could be used in the home for an older person living alone, or in a medical setting.

Perhaps the biggest contribution of social robots will take place *outside* a medical setting, in people's homes, where we make the decisions that directly impact our health and wellbeing. Imagine, for example, that you've been diagnosed with a chronic disease like heart failure, arthritis, or even cancer. You leave your doctor's office with a binder full of instructions and a prescription for a half-dozen or more medications—and you're sent home to fend for yourself. Even if you have a partner, the two of you may feel overwhelmed. You're frightened, maybe even confused by all the instructions. What should you eat? What kind of exercise is okay? Sure, you can turn to Google, but when you have in the past, you've found that the information online can be intimidating, and may not even apply to your case. And you (or your partner) can't call your doctor umpteen times a day with every question that pops into your head.

But you can ask Mabu, your home health companion, a social robot launched in 2014 by Catalia Health; the company's founder and CEO, Cory Kidd, PhD, was another of my MIT Media Lab colleagues. Mabu's mission is to help people better manage the challenges of chronic disease, like managing symptoms; dealing with stress, anxiety, and depression; sticking to a diet or exercise plan; and medication adherence through education and support. Its services are provided to patients free of charge by pharmaceutical companies and healthcare systems, such as Kaiser Permanente, to ensure the success of the treatment. Mabu has already worked with rheumatoid arthritis and kidney cancer patients. Since 2018, Catalia Health has been collaborating with the American Heart Association to use their treatment guidelines and educational content for heart failure patients.

The size of a small household appliance—think of a blender— Mabu is thoroughly trained in the user's specific condition. So, say for heart failure patients, it is loaded with material provided and vetted by the American Heart Association. It also draws upon information from other tools the patient may be using, such as a smart scale or a fitness tracker.

Let's say you're in a funk and you simply stop taking your medication, or you just sound "off." Mabu can send an alert to your health team. But Mabu isn't just a souped-up version of a health app that rattles off generic answers. It creates a true relationship between itself and the patient.

Designed by IDEO, the international design and consulting firm, Mabu, like Jibo, is very cute and approachable. It has a daffodil yellow face and body and big eyes (brown or blue) that blink. One camera allows it to make eye contact, and it can move its head to track the user's face. Mabu holds a tablet, so you can interact by touchscreen as well as voice. The social robot comes across as friendly, eager, and, well, *almost alive,* although not in a creepy way. There's something about the way it looks at you, and its soothing, feminine voice, that seems very comforting.

But Mabu's greatest strength is that the relationship between it and you is not purely transactional—it doesn't just send out automated reminders to take your meds, like so many other apps. Mabu can carry on a bona fide conversation that sounds like two people talking. In fact, Kidd employed a Hollywood screenwriter to help construct the dialogue so that the device would be both engaging and natural. The goal is to develop a warm bond between patient and robot health coach to ease some of the patient's burden in managing a chronic and complicated condition.

Through AI, Mabu develops empathy for the user and, using that empathy, devises the best way to interact with him or her. Like any relationship, this one develops over time. Literally right out of the box, Mabu will experiment with different approaches to see what works best with each user. For example, early on in the interaction, Mabu will typically crack a joke. If you seem uncomfortable or disapproving, it will shift to a more serious, no-nonsense mode. If, however, you enjoy the humor, Mabu will endeavor to keep things light. Over time, you and Mabu can establish a rapport.

The focus of healthcare in recent years has been on changing human behavior, because, after all, when it comes to health, it is the everyday decisions of individuals that matter most. And people don't always make decisions in their own best interest: According to the

CDC, one in five prescriptions for medication are never filled, and out of those that are filled, half are taken incorrectly, "particularly with regard to timing, dosage, frequency and duration." This has created an entire industry of tech tools, such as smartphone apps that alert you by text and pill holders with alarms to remind you to take your drugs.

As Kidd asserts, these tools often don't work because they're solving the wrong problem. "The reason that patients are not taking their medications has almost nothing to do with forgetting," he says. "Sure, that could be the case every once in a while, but that's not really the issue. There's a whole other set of things around education, around symptom management, side effect management, around stress and anxiety and depression, and those are the issues that we focus on."

When it comes to working as a home health coach, it's all about the *relationship*. If the mission is to persuade people to change their behavior—whether it's taking their medication properly, sleeping better, eating better, or becoming less sedentary—you must first build a bond of trust. Individuals must feel comfortable enough with the technology to share intimate information, ask questions, and most of all feel cared for and respected. And that is something that the healthcare system has not done very well.

Some studies have shown that patients in hospitals actually *prefer* dealing with avatars on a tablet to human beings, for the very reason that an avatar doesn't make them feel rushed or judged. They can ask the avatar the same question over and over again and not feel stupid, and at the same time not monopolize a doctor or nurse who needs to attend to a patient with a more serious problem. This is particularly true for mental health issues that could be potentially stigmatizing. A 2017 joint study by researchers at the Institute for Creative Technologies at UCLA and the Carnegie Mellon School of Computer Science found that military service members were more likely to be honest about their PTSD symptoms to an avatar (a virtual human interviewer) than to a real human interviewer. So, a Mabu user may be more willing to ask an "embarrassing" question of a robot than of a doctor or nurse.

Mabu is designed to work with individuals one on one in their homes, as a companion, support system, even buddy. It is not mobile; it doesn't need to be. There are other social robots, however, that are built to be out in the world, not just sitting on a table in a living room, but interacting with lots of people.

Launched by SoftBank Robotics in 2014, Pepper is a four-foot-tall interactive humanoid robot. Unlike Mabu, Pepper has an actual body of sorts that enables it to move around on three omnidirectional wheels. At HSBC's flagship retail branch on New York's Fifth Avenue, Pepper offers information on the bank's products and services and will summon a human if you want to apply for a mortgage, pose for a selfie, and show off some cool dance moves. At Washington, D.C.'s Smithsonian Institution, Pepper directs visitors to exhibits. Thanks to its efforts, foot traffic is up in some of the museum's lesser-known, underutilized exhibits. In San Francisco and Boston, Pepper is mingling with STEM students to teach programming in public schools through SoftBank Group and SoftBank Robotic's joint educational initiative. In Ostend, Belgium, Pepper greets visitors to the AZ Damiaan hospital.

There are some fifteen thousand Peppers distributed across the globe in retail malls, airports, offices, hotels, and on cruise ships. And more than a thousand are used in Japanese homes as companions.

If a robot is going to be out in the world mingling with humans, it has to have people smarts. Pepper is in testing to recognize basic human emotions like joy, anger, and surprise, and to enable it to alter its response based on the human's mood. In 2018, Affectiva partnered with Pepper's maker, SoftBank Robotics, to expand the robot's emotional capabilities, allowing it to better adapt its behavior based on a deeper understanding of people's complex emotional and cognitive states as it interacted with them. As we continue to develop our partnership, future versions of Pepper may understand more sophisticated emotions and feelings, such as drowsiness and distraction, and the difference between a smile and a smirk.

Designed in Paris, Pepper is appealing and nonthreatening in appearance—almost childlike, in a way that puts the users at ease. It uses a tablet embedded in its upper body along with speech to communicate with humans. With its grasping hands, Pepper can, among other things, gesture when speaking and wave. It speaks more than ten languages and can carry on a conversation reasonably well. Yes, at times it may falter and give a bizarre or goofy response (like Siri or Alexa), but considering everything else it does, it is a remarkable piece of machinery.

Pepper is also a people magnet. This may change when robots become more commonplace, but right now the robot has been proven to attract foot traffic, whether it's in banks, malls, or museums. Some people are simply curious about such robots, but others may be more comfortable starting a conversation with a robot than a human being. "This presents a new opportunity for retailers," explains Matt Willis, PhD, design and human–robot interaction (HRI) strategy lead at SoftBank Robotics America. "Somebody might not be ready to talk to a salesperson, but they may be willing to ask a robot for information. After that, they may feel ready to talk to a human employee and by which point, they will be able to have a more informed conversation with that sales associate. We're providing value both to the end customer and also to the store or to a sales associate by having a robot in that loop."

Eventually, robots like Pepper could have an expanded role in retailing. For example, with access to a store's database, a robot could not only answer a customer's questions about a product, but also make sure that the product is in stock in the right size and direct the customer to its precise location. Or, if someone is coming in to pick up an online order, a robot could direct that person to the right department and, at the same time, suggest other products related to the purchase, just as if the customer were shopping on Amazon.

"In a sense, this approach is bringing the digital experience into the physical world," Willis notes. "Having a humanoid robot in the retail space means that I can have some of the benefits of the digital world right there in my retail store."

I believe that the merger of digital and human beings (shoppers and salespeople) in the physical world is the future of retail, in a way that will seem very natural, especially for a generation accustomed to conducting so much of their lives online. It's not a matter of if but *when* robots will be interacting with human beings in malls, offices, hospitals, banks, museums, and airports. But these machines must have a basic knowledge of "human" to work and live among us. And we humans will need some knowledge of "robot" to keep things working smoothly.

That's one reason Willis feels that SoftBank Robotic's educational initiative, providing Peppers for use in classrooms in San Francisco and Boston, is so important. "If we take it as fact that there are going to be more robots in the world, and they will become more and more capable to do things over time, then we also need more and more people to be interested in and able to work with them," Willis says. "Our educational initiative is not just about teaching computer science; it's also training the workforce of the future."

I began this chapter with the bad news about Jibo, but let me end on a more upbeat note. My Jibo and a few hundred others have been saved. They have been enlisted in a research project at MIT exploring human–robot relationships, and will now live on MIT servers.

There may be some bumps in the road (like Jibo), but I believe it is inevitable that social robots will become part of our everyday lives in the near future, just like our smartphones. They will be so embedded in our everyday activities that we often won't even notice that they're there. We will just turn to them when we need them.

As Jibo said when he bid us goodbye, "Maybe someday, when robots are way more advanced than they are today and everyone has one in their home, you can tell them that I said hello."

25

Alexa, We Need to Talk

As Affectiva's CEO I am now the official "face" of the company. I do a lot more public speaking, which I really enjoy, but I never take for granted that I will do well: I still practice right up until the presentation. One morning at home, while I was rehearsing a speech out loud, I happened to mention "Amazon Alexa." This woke up my Alexa. "Playing Selena Gomez," she announced from across the room. A second later, she began streaming Gomez's hit song "Come and Get It."

Except I hadn't asked for that song, and I didn't want to hear it. I had to yell, "Alexa, stop!" several times before Alexa shut it off. Although I was getting frustrated, Alexa, like the majority of virtual assistants, was completely oblivious to my feelings.

At that moment, I was annoyed. I didn't expect Alexa to be perfect—even human beings make mistakes—but she could at least have acknowledged that she'd made a mistake and responded, "Oh, sorry, Rana. I see that I misunderstood you."

This wasn't all that different from my experience years before, when I was a doctoral student at the Cambridge Computer Lab, practically living on a laptop that couldn't recognize or respond to the fact that I was homesick and lonely. It made me feel that I was being ignored, not being listened to. That was just how I felt now.

With our technology so embedded in our lives, and our interactions with them practically nonstop throughout the day, we expect more of them. We treat these devices more like partners than machines, and we expect them to behave accordingly. When they don't, when they seem out of sync with our intentions, it's jarring.

My team and I were moving Affectiva in a new and broader direction at the time, with the goal of becoming the leader in Emotion AI. That meant that we as a company had to think of the big picture: What was the future of the human-to-computer interface?

Conversational interfaces were becoming more and more mainstream. What's easier than talking to a device? No keyboards or screens to fuss with. No learning curve. When it works, it's effortless, and that's what people want from their devices.

Launched in 2014, Amazon's Alexa was already in millions of homes. Siri had been around since 2011. Google Home had just been released, and Samsung was investing in Bixby, its version of a conversational interface.

I knew it was just the beginning. More and more, we would be interacting with robots and virtual systems that, like human beings, could see and hear (through computer vision and machine hearing). This meant that as a company, we would have to start thinking more broadly. If we wanted to be the company that provided a holistic view of how people felt—a multifaceted Emotion AI company—then we would need to add a voice component.

We had always recognized that the face is just one channel of communicating emotions—a critical one, to be sure (and my favorite), but certainly not the only one. Voice and body language are also important; some would argue, of equal importance. The reality is that, depending on the situation, humans seamlessly switch among all these modes of communication. When you're talking on the phone, you put more emphasis on communicating emotion with your voice than you would if you were speaking with someone face-to-face. Or if you're standing across a room trying to attract someone's attention, you're going to use big gestures to get your message across. Sometimes the ways we communicate contradict one another; people who are very frustrated often smile. It's a wry smile, not a joyful smile—but without context, it can be confusing.

When I speak to my mom by phone, even though we are half a world apart, I know within seconds, just by how she says hello, whether she is in a good mood or something is bothering her. And within a few seconds of a call, if I sound down, she'll pick up on it. "Rana, is everything okay?" The prosodic features of our voices—that is, the pitch and intonation—convey the emotion and cognitive state behind the words and facial expressions. These include things like how loud you speak, how much energy is in your voice, or even how fast you speak. Just as we are hardwired to interpret facial expressions, we are also ingrained with the ability to decode vocal intonations.

Affectiva would, therefore, have to teach a computer to decipher vocal intonations the way we had trained it to decode facial expressions. And it was pure serendipity that, just when we decided to build a speech team, the following message popped up on LinkedIn from a speech scientist. "Hello, Rana. The work you are doing at Affectiva is very inspiring . . . Several of my research projects have dealt with building statistical models using prosodic features for emotion. My goal for future work is to combine speech, facial gestures, and hand motions to estimate emotion in a vein similar to what you are doing at Affectiva. May I send you my CV?"

I said yes, of course. And today, Taniya Mishra, PhD, is speech scientist and director of AI research at Affectiva. Taniya's credentials are spectacular, and I was eager to mentor a woman in AI, a field where we are underrepresented in top positions.

Although we come from different countries and cultures, Taniya (pronounced like Tanya) and I have a great deal in common. Like me, she is a mother juggling work and family life. Born and raised in Calcutta, India, she is the daughter of two physicians who, like my parents, prize education and service to the community. Perhaps it was the fact that India recognizes twenty-two official languages that sparked Taniya's interest in the spoken word. At a very young age, she was fluent in four Indian languages, and much to her parents' surprise, she also spoke English.

"That was my parents' secret language," she told me. "They spoke to each other in English when they didn't want me to understand what they were saying, but I just picked it up."

Just as I became intrigued with faces at an early age, Taniya became fascinated with voices. As she puts it, "Think of voice as a layered cake—our voices can tell people our gender, age, and even personality. You can tell if the language I'm speaking is my first or my second language. It conveys your emotions, it conveys your cognitive state."

Taniya got a PhD in computer science from Oregon Science and Health University. Her thesis project was developing emotional speech synthesis systems for children on the autism spectrum; in other words, she was doing with the voice what I was doing with the face, creating a tool that would help autistic children decipher emotion.

Most of the voice analytic technology today requires a full sentence or at least several words before the algorithm can analyze it. Taniya invented a system that can analyze emotion from speech in real time, second by second, exactly as I did for the face.

This is critical because emotion is something that plays out over time. As Taniya explains, "Take anger, for example. People don't usually go from being calm to angry in a second. Usually there's a build-up to it. You go from being neutral, to being a little annoyed, then being a little frustrated, then being angry, and then raging. It's like a spectrum."

The goal in terms of Emotion AI is to detect the rise to anger as early as possible, so that someone (in this case, something like an

Amazon Alexa) can intervene before you become furious. Many call service centers use voice analytics to track customer sentiment, so that service representatives can respond appropriately to customers who may be getting increasingly unhappy or frustrated. This was one of the earliest uses for this technology. But as with facial Emotion AI, there are almost limitless possibilities for voice analytics.

Today, Affectiva is partnering with two market research firms that use voice and face: Living Lens and VoxPopMe. The combination of the two channels, voice and face, is more powerful than either one alone.

There is also a tremendous amount of research being done in healthcare to create vocal biomarkers for different mental and physical ailments—diagnostic tools that can detect changes in vocal tone that may indicate current problems or problems down the road. For example, using a smartphone app developed by Tel Aviv–based Beyond Verbal, researchers at the Mayo Clinic, in Rochester, Minnesota, found abnormal voice characteristics in patients with coronary artery disease, compared with those who were disease-free. Imagine if simply tracking vocal changes over time could one day be used as a diagnostic tool to detect heart disease.

Voice analytics has potential applications in mental health, too. Numerous mental health research groups are exploring the use of vocal tracking on smartphones or home robots to check the progress of patients and look for signs of depression or suicidal intent. After all, a lot of us today talk to our devices even more than we talk to our friends, family members, or healthcare professionals!

The training model for voice analytics is similar to that for face—except, of course, you have to feed the algorithm lots of samples of people speaking. Affectiva's approach to voice is language agnostic: The algorithm is listening not to the spoken words per se, but rather, to the vocal utterances—that is, how those words are being said.

Our voice team's first project was to build a real-time anger and laughter emotion detector. Because we had to feed the algorithm an abundance of data, we turned to databases of customer-service calls.

Have you ever become angry when complaining to a company's customer service rep about a faulty product or a mistake on a credit card bill? Well, more often than not, you're informed at the beginning of the call that it is being recorded for "quality and training purposes." Sometimes those recordings are "scrubbed" of the person's identity and used for research databases. You don't know who is talking (or screaming), but you can hear individual callers yelling and complaining. These recordings provide wonderful data sets of real people in real stages of angst, anger, and frustration.

We also fed our algorithm English, German, and Chinese utterances so that it would have a sample of different spoken languages. Again, the software was not listening to the words. It was listening to how fast and how loud the person was speaking, whether she was speaking in a monotone or whether she was excited, as demonstrated by variation in pitch and intonation.

Interestingly, the algorithm listens for the silences *between* words and sentences as well; pauses are very important for interpreting the meaning of words. For instance, if I say, quickly, "Yeah, I love that," it means something entirely different from what it means if I pause and say, more slowly and deliberately, "Yeah . . . I love that." When you pause, it's a sign that you're not really persuaded yet; you're still thinking it over. That's a powerful piece of information.

Similar to our work with the face, we had human labelers, or annotators, listen to the voices and categorize them as happy or angry.

When it comes to our interactions with technology, frustration can play a major role in how we respond to our virtual experiences. If a consumer's frustration goes unchecked, it can harm a company's reputation and brand, and cost it sales. As the 2018 Customer and Product Experience 360 survey put it, "The Smart Home Is Creating Frustrated Consumers: More than 1 in 3 US Adults Experience Issues Setting Up or Operating a Connected Device." Based on this study, 22 percent of consumers simply return the product.

Let's say you're talking to a virtual assistant in your car, like Alexa. Perhaps you're asking Alexa to play your favorite song, or compose or send text messages to your spouse or partner, or get someone on the phone for you. If your virtual assistant stumbles, you get more and more annoyed. Your frustration builds, and you can get distracted, which puts you at greater risk for having an accident. That's just one reason it's important for your virtual assistant to understand when you're feeling frustrated.

The first step to deescalating such frustration is to understand it. That was another major research project undertaken by our voice team. But this time, we looked at both face and voice.

One of my first projects undertaken as CEO was to pursue opportunities in the auto industry to bring emotional intelligence to our cars. That meant being able to track the emotional and cognitive states of drivers. To conduct our research, we built an automobile simulator (a smart dashboard) in our lab and used it to conduct a study on vehicle tech–induced frustration.

We brought people into our lab and asked them to interface with Alexa—but in this case, we had modified Alexa so that it kept getting whatever the person requested wrong, over and over again. The name for an experiment that involves a human interfacing with a system that the participant believes is autonomous but is in fact rigged is "a Wizard of Oz experiment."

The purpose of the study was to purposely induce frustration while at the same time collect data on the participants' face and vocal expressions. We wanted to find out which was the better way to detect frustration, voice or face.

We re-created the kinds of situations drivers encounter every day (though anyone who uses a virtual assistant or social robot encounters the same kinds of interactions). The participants were instructed to ask Alexa to add items to or remove items from a shopping list; tell a joke; set timers; play various radio stations; turn on a song, audiobook, or the news; or compose and send text messages. We tinkered with Alexa so that her performance was abysmal, deliberately designed to rev up the participants' frustration levels.

From this study, we learned that you can't accurately detect frustration from either voice alone or face alone; you need input from both. From our work, we developed training models for our algorithm. It was the first time that anyone had combined facial expressions and voice to create a frustration detector. We think of it as an antidote to technology-induced frustration.

In the near future, I suspect that most of our consumer technology will be equipped with a frustration detector and protocols for how to defuse frustration early, before it boils over. This will be especially important for the tools we use in everyday life.

Amazon Alexa's AI team has gone on record saying that they are experimenting with ways to allow Alexa to detect emotions like happiness, sadness, and frustration. In fact, the company has come up with the Alexa Prize Socialbot Grand Challenge 3, "a multimillion-dollar university challenge to advance human–computer interaction" that encourages researchers to create socialbots that "can converse coherently and engagingly with humans on a range of current events and popular topics such as entertainment, sports, politics, technology, and fashion."

The reality, however, is that until Amazon and other virtual assistants and social robots have incorporated Emotion AI, humans' relationship with these technologies will remain transactional at best—not truly collaborative. Our interactions with these smart devices will feel clunky and machinelike until the devices can "converse" and respond to us in a more natural way.

Perhaps in the near future—maybe the next time I'm rehearsing a talk on AI in my living room—my upgraded Alexa won't mistakenly play some song I don't want, but instead will interject, "Good point, Rana. I think you nailed it. I really liked your description of me as being 'emotionally attuned to the needs of the user.' I certainly try hard. Just one piece of advice: I think you could be a bit more assertive at the end."

26

Robots on Wheels

I n 2012, while I was "grounded" in Cairo, trying to fix my marriage, I had driven nearly an hour from my New Cairo home to visit my father-in-law, Uncle Ahmed, in his office. He offered me some fatherly advice that day: Forget about work (which by now I was doing mostly remotely from Cairo) and focus more on my "wifely duties," like cooking. I trusted Uncle Ahmed and knew how badly he wanted Wael and me to reconcile, but I had my doubts that my efforts were going to pay off. I was sobbing when I left his office, but determined to follow my new marching orders.

The traffic was light on the drive home, a straight shot back to the suburb where I lived. I was going through all the motions of driving, but my brain was elsewhere. I kept replaying what Uncle Ahmed had told me. I started planning a fabulous dinner for that evening and then realized that I was missing a few key ingredients. Impulsively, I reached into my bag on the passenger seat, grabbed my cellphone, and dialed Gourmet Egypt, my neighborhood grocery store, to place an order.

Whack! A second later, the phone flew out of my hand as I slammed into the steering wheel of my Volvo.

I had plowed into a truck at sixty miles an hour when its driver cut me off making a left turn across the road. My car was totaled. A few passing drivers stopped to help. I was dazed but conscious, able to walk but badly shaken up. When I think of how much damage my car sustained, I know that I got off easy. I walked away with a few bruises and a dislocated shoulder. As we say in Egypt, *Rabena Satar* (God saved me). It could have been much worse.

Honestly, whatever the other driver did, I had been in no condition to be behind the wheel of a car. Given my knowledge of how your emotional state influences your decision making, I should have known better. But I hadn't been thinking straight. *My car should have known better,* I thought. *My car should have had my back.* It should have anticipated when I got in it that I was an accident waiting to happen and intervened before I posed a danger to myself and others on the road.

Right now, conventional cars know very little about the occupants in the vehicle. They don't know who is driving, what the passengers are doing, or the state of the driver's emotions. Is a teenager driving who is texting on the brink of losing control? Are a sister and brother in the back seat fighting over a toy and causing a parent to turn around in anger to reprimand them? The vehicle doesn't even know how many occupants it is carrying, and this information can be critical. Every year in the United States, on average, thirty-eight young children die of heatstroke because they are inadvertently locked in hot cars by harried or distracted parents and caregivers.

Nor does a conventional car know whether the person behind the wheel is too emotionally distraught to drive safely, as in my case. I was an accident waiting to happen, and unfortunately, my story is not uncommon.

According to the CDC, human error is responsible for 94 percent of all serious auto accidents. Distracted driving (texting, phone calls), drowsiness, alcohol impairment, road rage, speeding, and poor judgment calls are the reasons for most of the 1.4 million auto-related deaths worldwide each year. More than forty thousand people in the United States die in automobile accidents every year. (That's as many as the number of women who die of breast cancer each year, a major public health crisis.) Cars are the eighth leading cause of death in the world.

The solution to human error seems obvious: reduce (or eliminate) the role of the human behind the wheel. Toward that end, for decades car companies have been packing more and more IQ in their cars, software that automates some of the tasks of driving. These include lane departure sensors and drift control, which keep you from inadvertently veering into another lane, and virtual assistants that can route your trip so that you're not fidgeting with a map. Semi-autonomous driving features (like the autopilot offered by Tesla and Cadillac) enable a car to drive itself for a portion of the ride to give you respite. Blind spot detectors, vehicle backup cameras, self-parking features, and automatic braking are all designed to keep us safer. These features focus on the environment *outside* the car, but they tell the car virtually nothing about the people *inside*.

Even with all these safety upgrades, auto-related deaths in the United States haven't declined. They've actually edged up in recent years. Why? One big reason is our smartphone addiction: Some people never put theirs down, even when they're driving or crossing the street. Ultimately, it's the responsibility of the driver to anticipate these and other unexpected events, but drivers can't do this quickly enough if they are texting, talking, or otherwise distracted. That's why knowing what's going on inside the car is equally important to knowing what's going on outside.

Without greater knowledge of the human being behind the wheel, conventional cars are ill-equipped to deal with human behaviors that are the primary cause of fatal accidents. In short, understanding "human" is as much a part of auto safety as seatbelts and air bags.

In light of my experience, creating Emotion AI tools for auto safety has been a priority for my company and me. Shortly after I became CEO at Affectiva, a Japanese automotive manufacturer challenged us to adapt our technology for an automobile. I wasn't surprised by the request: When I was at MIT, our automotive sponsors were among the most interested in our technology.

Automobile companies were curious about how people *used* their cars: What activities did the driver and passengers engage in? It was important for them to know whether the driver and passengers were having an enjoyable and comfortable experience, or whether they found operating the vehicle to be annoying or confusing. Ultimately, the auto manufacturers wanted to know how people *felt* about their ride, in real time, in the real world.

Gathering all this data was a monumental task. For starters, we had to get our technology to work under a variety of conditions that we didn't have to consider for the purposes of market research. For example, the system had to function in the dark at night. That meant we had to retrain our algorithms to handle videos that were collected using near-infrared night-vision cameras.

Then we had to teach the technology to track a face even if the drivers or passengers were wearing something that partially blocked their faces, like sunglasses, or face masks to protect against germs, a common practice in Asia, especially in an enclosed environment like a car. The algorithm also had to know if someone was looking down and texting, or eating a hamburger while driving. (We call all these situations "occlusion" because when they are happening, parts of the face are occluded, or blocked.) To train the algorithm to do all these things meant we had to collect a massive amount of data.

We enlisted volunteers to install cameras in their cars, and we collected data while they were on their daily commutes—with their consent. We were shocked by some of the behavior we saw behind the wheel: A father fell asleep while driving his toddler to daycare and woke up seconds before he crashed into another car. A woman who was driving while texting was using not one but *two* phones simultaneously, and barely glancing at the road. And then there was the unforgettable carload of inebriated teenagers passing around a bottle of whiskey late at night! These scenarios drove home the message that our roads are unsafe, and I knew our technology could help change that.

This research produced Affectiva's Mobile Lab, a hot-pink Honda loaded with Emotion AI tech that simultaneously tracks what is going on both inside and outside the car. If you live in Boston, you may have spotted it cruising around town or parked in front of our office.

The Mobile Lab has an outward-facing camera to observe what is happening outside the vehicle and an inward-facing camera attached to the rearview mirror that has a full view of the interior, to track both the driver and the passengers. Microphones pick up voice intonations (not spoken words, but the sound and tone of the voice). Using the data collected from the camera and microphones, the Mobile Lab can unobtrusively monitor and react to the mood and cognitive states of the driver and passengers and observe what they are doing (such as whether the driver is eating a sandwich, texting, or being distracted by a passenger).

This technology equips vehicles with common sense (at times, more than the humans in the car possess). So, like the friend who doesn't let a friend drive drunk, the smart machine can intervene, and even take over, when human beings are on the brink of making a potentially fatal mistake.

I often relive my accident in my head, and imagine how differently it would have unfolded in an emotion-enabled vehicle like the Mobile Lab, starting from the very first moment I got behind the steering wheel. An emotionally intelligent car would have "seen" the tears rolling down my cheeks and my eyes, puffy and red from crying, and determined that I was upset. From that data, it would have understood that I was likely to be less attentive as a driver and would automatically have shifted into "high alert."

Through the conversational interface in the car, it might have said in an empathetic voice, "Hi, Rana. I see that you are upset and are having a bad day. I am so sorry you are feeling this way. This puts you at a higher risk of getting into an accident. Please stay safe. I will also do my part and be extra vigilant."

This gentle warning might have been enough to remind me to stay focused on the road. But my car would also have been keeping an eye out for erratic behavior. As I drove, my brain racing with a million thoughts other than what lay ahead on the road, it would have been tracking my head movements, watching my eye gaze patterns—people who are mentally distracted do less gaze shifting and have more tunnel vision. Based on my actions, my car could have recognized that I was mentally distracted, oblivious to the road ahead of me.

When I reached into my bag to pull out my phone, my car would have observed that my hands were off the steering wheel, my eyes off the road, and it would have sprung into action. Via its external cameras, it would have seen the truck about to cut me off and, based on my eye gaze patterns, would have determined instantly that *I* did *not* see it. At that point, it would have shifted its controls to autopilot and taken over the braking function. And the crash would have been averted. I would have been jolted back to reality, unharmed. The car then would have recognized that I was once again capable of driving myself home and restored the driving function back to me.

There isn't a commercially available car yet today that has the people smarts to do what my Emotion AI–endowed car could have done to prevent my crash. But that's about to change.

ROAD TO AUTONOMOUS

Within the next five years, sophisticated Emotion AI will be offered in new car models, equipping the cars with "intuition" to understand and respond to the cognitive and emotional states of the driver and passengers. Why? Car companies are quickly learning that the more IQ you pack into an automobile, the more EQ it needs.

The next big thing in the automobile industry is "autonomous," or self-driving, cars—cars that are autonomous if not all the time, then at least part of the time. Semi-autonomous vehicles in their first iteration will be about not eliminating human drivers, but reducing human error. These cars will be built not to run on their own, but to be *co-driven* with a human driver. That is very different from driverless vehicles whisking people around town and, in essence, rendering drivers obsolete.

As smart as AI is today, it is not infallible, at least not yet. No matter how well you train an algorithm, situations will arise that it doesn't understand. In literally the blink of an eye, the car may suddenly relinquish control back to the driver, as the driver may see a problem that the software does not recognize. In these cases, human drivers must be prepared to take over.

We call this interaction between the driver and the smart car "the hand-off," and it is one of the primary reasons that the move to so-called autonomous cars will only accelerate the demand for emotionally intelligent vehicles. A semi-autonomous vehicle cannot operate safely unless the driver and the machine are in sync with each other. The car can't hand off the driving function to a driver who is distracted or dozing, texting or impaired. Fatalities in semi-autonomous cars are rare, but when they do occur, it almost always involves a glitch in the machine-to-human hand-off (i.e., when the human driver is distracted and doesn't intervene when he or she should). Even when the smart car drives like a dream, the human driver must be focused enough on the road in case something goes awry and he or she must intervene.

AI is sophisticated enough to take over the basic function of driving a vehicle, but driving entails so much more than just navigating a car. Human beings perform a vast variety of tasks in a car, and much of that job is dealing with other humans. Look at the role of the driver: Only a part of his or her attention is focused on actually operating the vehicle. There are also ongoing interactions with others in the car— "Uh-oh, Michael is carsick and about to throw up! Can you turn down the music, heat, or air-conditioner?" "Hey, you two kids had better stop fighting or I'm going to pull over." Or, for example, if another car is about to cut you off, your passengers will automatically check to see if you are aware of the situation and are prepared to react.

A driver also often watches and signals to people and other vehicles outside the car. Imagine, for example, that you're stopped at a crosswalk, about to make a left turn, and you see a distracted pedestrian stepping into the crosswalk. You don't just plow ahead. You observe her body language; you make eye contact. If she nods and gestures for you to go, you might proceed with caution, or you might wave her through.

The fact is, human beings won't completely be out of the driver's seat anytime soon. We still have the edge over AI because we *think* like other human beings. We anticipate that other drivers or pedestrians may not always follow the rules of the road, or they may behave in irrational ways. If we see children playing ball near the road, we will mentally take note and be prepared to stop if a child runs into the street to retrieve the ball. We know this not from a driving manual, but from life experience. We instinctually respond to situations in a very human way. Eventually, learning algorithms will be developed to imitate some of our gut reactions, but this will take time.

So far, there's been little effort on the part of either the automobile industry or public safety officials to bring the public up to speed. When people hear the term *autonomous car,* they are often under the mistaken belief that the role of the human driver has been eliminated, that he or she is now relegated to the role of passenger.

"The biggest myth about automation is the more automation, the less you need human expertise. Actually, the more you automate, the more you need to educate, where, when, how, etc.," observes Bryan Reimer, PhD, a research scientist at MIT's Center for Transportation and Logistics, a researcher in the AgeLab, and associate director of the New England University Transportation Center.

Dr. Reimer, who has studied driver behavior as it relates to automation, raises an interesting point: As our automobiles assume more and more of the tasks of their operation, there is a risk that human beings will lose some of the real-world driving experiences that make us seasoned drivers. This is fine as long as the car remains in control. But in a world of semi-autonomous vehicles, the drivers may become less equipped to handle the complicated maneuvers that would stump the software. As Reimer notes, "We unfortunately will get worse at driving, as humans learn from *doing*. The less we do, the less we learn. That's why a lot of the risks of this mixed system go up over time. If we're no longer doing, we're no longer learning. So, that means the future is one of novice drivers, and we all know that novice drivers are even more risky than established, trained drivers."

So, there is a catch-22 here. If we lose our driving skills due to semi-automation, we won't be able to be as effective partners with our semi-autonomous cars. Furthermore, the very nature of semi-autonomous vehicles may lull some people into a false sense of security.

In addition, when we get behind the wheel of a car equipped with lots of sophisticated automated features, human beings revert to behavior that can actually make us less safe. It has to do with what psychologists call *cognitive load,* a field of research that began with a study of mice in a maze. In 1908, psychologists Robert Yerkes and John Dillingham Dodson found that mice given a low electric shock were motivated to finish a maze, but if the shock was too high, the mice gave up. This led to the famous Yerkes-Dodson law we still refer to today; its bell-shaped curve shows the relationship between arousal and performance.

When you move to the left of the curve, you are less aroused, and when you move to the right of the curve, you are overaroused. Finding the right amount of stress is tricky: If stress levels rise too high, it can result in *cognitive overload,* and people, like mice, shut down. If, however, stress levels dip below a certain point, there is a risk of *cognitive underload,* which can also impair human performance. For optimal performance you need to land on the just right spot on the curve. But maintaining the right level of arousal in a car can be tricky.

For decades, it was assumed that automation induced underload (a drift too much to the left of the Yerkes-Dodson curve) and would make people drowsy or less alert. In some instances, that was true. In the late twentieth century, researchers noticed that this was exactly what was happening to pilots in cockpits as the once-labor-intensive tasks of flying a plane were replaced with computerized navigation and safety equipment. But cognitive underload in an automated car looks different from, say, that in a cockpit. When we are understimulated and veer to the left side of the bell curve, we often become bored and look for some other, nondriving activity to occupy our time, such as texting, making a call, eating, or watching a video.

"And that's where the fundamental premise of state management comes in, helping people to make better moment-to-moment decisions," Reimer explains.

Car companies are beginning to add features to keep drivers more fully engaged during periods when the car is in control. General Motors's 2018 Cadillac CT6 offers a Super Cruise system that enables "the first true hands free driving system for the freeway," on some U.S. highways, but it has a camera embedded in the steering column that tracks the driver's head position and eye movements to make sure that his eyes are on the road.

The ProPILOT Assist system offered on the Infiniti QX50 and Nissan Leaf takes a different approach to keep drivers engaged: In order to operate it, the driver must keep both hands on the wheel. These features help, but as more and more Emotion AI tools are embedded in cars, there will be more sophisticated ways to gauge human attention and mood and to keep the driver alert and ready to take the wheel.

Bottom line, it doesn't look like we're going to have to shut down our driving schools anytime soon. But the way we approach driver training will likely need an upgrade to accommodate the new world of semi-autonomous cars. Perhaps we will need to periodically take refresher classes to maintain our driving skills, or practice on tracks that are filled with the kinds of unexpected challenges that occur in real life.

We could be in what I think of as a semi-autonomous transition period for quite a while—decades, in fact. Still, there will be sectors that move quickly into the fully autonomous zone. Right now, there are pilot programs worldwide (run by Uber, Lyft, and Waymo) to test self-driving robocabs.

Affectiva recently partnered with Aptiv, a leader in developing in-car sensing and autonomous vehicles. In 2017, Aptiv acquired Boston-based nuTonomy, founded by Karl Iagnemma, PhD, former director of MITs Robotic Mobility Group. In 2016, nuTonomy offered "self-driving" taxis to Singapore residents. The company has piloted an autonomous vehicle in Boston and, under Aptiv, is partnering with Lyft to offer self-driving taxis on the Las Vegas Strip. The cars are autonomous in that they can assume all the functions of a vehicle, but they still operate with "guides" in the front seat. The guides are ready to assume control if needed and soothe the nerves of skittish passengers who may be curious about traveling in an autonomous vehicle but remain wary.

So, what will it take to make people feel so confident in a robot on wheels that they're willing to trust their cars to drive them? Dr. Iagnemma, now president of Aptiv Autonomous Mobility, says the plan is for the taxis to be completely driverless in some locations soon. He admits, though, that how fast the company moves to complete autonomy hinges in large part on how consumers feel about riding in a car driven by an AI system.

Iagnemma says the company has to develop a car not only that is safe, but also that the passengers *believe* is safe, "and those are two totally different things. That gets to the emotion context, among other things. It also gets to subjective things, like how does the ride feel, the quality of the experience, those sorts of things.

"You could wrap it up in a bundle by saying there has to be an *emotional response* to that system that is positive, that will result in someone, when they get out of the car after a trip, wanting to take another trip in the car. Otherwise, the technology becomes one of those things that has tremendous promise but ultimately doesn't get adopted. And that would be a shame."

I figured that I spend about thirty-three days a year driving my kids around to their extracurricular activities. That's a lot of time driving a car, and it doesn't include the amount of time I spend commuting to work every day. I still want to spend the time with my kids, but I would much rather be fully engaged with them while we're traveling than focused on my driving. If I didn't have to drive, we could be catching up with each other, watching the news, reviewing homework, putting the time I would be spending behind the wheel to better use. But I would have to be *very* confident in the technology before I felt comfortable doing this. I'll admit I'm not there yet, but I think I will eventually get there.

The average American spends close to an hour every day in a car commuting to and from work. Imagine what your commute would be like if you didn't have to drive, if, instead, you got in your car in the morning and it didn't look anything like the interior of cars today. With an AI in charge, it wouldn't even need a steering wheel! It wouldn't need traditional passenger seats. Instead, the interior would be designed for multiple purposes. This morning, you'd select the office mode. You'd take out your laptop, sit in a desk chair (with a seatbelt), and spend a half hour or so preparing for the day, doing work and making calls.

At night, you'd select the spa mode. The desk would turn into a side table, the office chair would be set in lounge mode. You'd sink into the chair and put up your feet. Sensing your fatigue, the AI system would lower the lights and release a calming lavender scent into the air. You'd close your eyes and listen to relaxing music, or the sound of a waterfall, or birdsong, and doze for half an hour. Or you'd read an engrossing novel, a memoir. You'd have your virtual assistant arrange to have sushi waiting at your doorstep when you got home. When you pulled up to your house, you'd feel refreshed, ready to spend time with your family.

Isn't that better than sitting in traffic?

27

Human Before Artificial

As AI becomes mainstream and ubiquitous, AI systems that are designed to engage with humans will have a lot of data on *you*—personal data about who you are, your preferences, your actions and your quirks. We live in a society where data is being collected on all of us, all the time. Sometimes it's obvious; sometimes it's not. Sometimes it's for our benefit; sometimes not. As the CEO of an AI company, I understand that having access to so much data comes with a great deal of responsibility: How do we ensure the ethical development and deployment of AI?

My team and I recognize that Emotion AI technology knows lots about you: your emotions, your facial and vocal expressions. It will know when you deviate from your norm and when you are having a bad day. We respect that this is very personal data, and we respect people's privacy. The fact that we deal with such highly sensitive data has heightened our awareness of privacy and consent.

We are passionate about the ethical development and deployment of AI. Wherever our technology is deployed, we always ask for opt-in and consent, and all the data in our repository has been collected with clear consent. That means we don't collect data from people who are not consciously aware that we are doing so. To date, we have collected around nine million faces from eighty-seven countries, all of which were recorded with opt-in and consent. Moreover, we also have adopted the European General Data Protection Regulation (GDPR), the toughest consumer data protection standard to date, which provides that consumers can ask that all their data be deleted from our database. In fact, in industries such as automotive and social robotics, we do not even store the data. The technology runs "on the edge"—for example, in the case of cars, on the electronic chips in the vehicle; it is not sent up to the cloud for analysis. Whatever happens inside the car stays inside the car; no data is recorded or stored.

PRIORITIZING PRIVACY

In this book, I cite many positive uses of Emotion AI, such as our work with the autistic community, in mental health, in creating new biomarkers for disease, in detecting distracted driving, in democratizing education, and in eliminating unconscious bias in hiring practices. My team and I offer a positive vision for AI, but we're not naïve. We understand that the potential for abuse, especially the careless management of data, is very real and it is happening right now, by companies and governments.

Privacy advocates have long voiced concerns over how data is collected by Big Tech, and how it can be sold or hacked, or can inadvertently end up in places where it doesn't belong. Technology companies have a long history of acting first and asking for forgiveness later. But lately, the public is not so forgiving. Facebook faced steep fines and significant reputational damage after Cambridge Analytica, a British political consulting group, harvested the data of eighty-seven million Facebook users without their consent and targeted them with political ads throughout the 2016 election cycle. People finally woke up to the fact that maybe all this data they were offering up was ending up in places they didn't approve of.

It's not just Facebook. Let's face it, every time we search the Internet, shop, or download a book or video, we are being watched. Like it or not, a lot of people whom we don't know know an awful lot about us and our preferences.

Stalking you online while you surf your favorite stores to learn your brand preferences (without your permission) may be an infringement of your privacy, but it pales in comparison to how totalitarian regimes have weaponized AI against entire groups of people. In April 2019, the *New York Times* reported that China was "using a vast, secret system of advanced facial recognition technology to track and control the Uighurs," a predominantly Muslim ethnic minority population. China may be the first country, as the *New York Times* says, to use AI for racial profiling, but there's no doubt others will follow.

Affectiva has very strong core values regarding the need to protect privacy and avoid situations like those I have just described, which are unethical or cause harm. On ethical grounds, we turned down forty million dollars from an investment group that was an arm of a government agency involved in surveillance. On principle, we avoid selling our technology to security and surveillance companies when there is no opt-in and consent. We can't, of course, control what our competitors do, but as a company, we can set our own standards and take a leadership position to push for the best practices the AI industry should adopt. For example, Affectiva is part of the Partnership on AI to Benefit People and Society (PAI), a tech industry consortium established to "study and formulate best practices on AI technologies, to advance the public's understanding of AI, and to serve as an open platform for discussion and engagement about AI and its influences on people and society." Its eighty-plus members from more than thirteen countries include tech giants as well as other voices, like the American Civil Liberties Union, Amnesty International, and the Hastings Center, a bioethics research institute. Affectiva is one of the few start-ups invited to join PAI, and within it, we are part of a working group committed to developing best practices for Fairness, Accountability, Transparency, and Ethics (FATE) in AI. It is encouraging to see players in the tech industry start to take positive steps toward tackling some of these issues. A consortium consisting of a diverse group of interested parties is the right approach. This issue must be tackled by a variety of stakeholders, organizations, and decision makers working together.

Unfortunately, not all AI companies or even sovereignties share the same core values around data consent and privacy. China poses a particular challenge for an independent AI company like ours. AI companies in China, like other businesses there, work closely with the government. They are often funded by the government, and they have access to the massive amounts of data on people collected by the government in ways that a democratic society would not tolerate. This gives them an edge in every way: They can crank out technology quickly, they can expand rapidly, and they don't have to consider ethical issues in doing so.

China's stated goal is to be the world leader in AI by 2030, and to do whatever it needs to do to achieve that. So, how does a company like Affectiva compete with its Chinese competitors? We can't compete—at least not on their playing field with their rules—so we must change the game. We must force them into a different competitive arena, one where we have a decided edge.

Totalitarian regimes ignore the public because they can. But in a global marketplace, the "public" isn't just your own citizens. You have to consider the citizens of the world. We can build a better AI on our strengths and ethical standards, one that empowers individuals. At the same time, we can reform Big Tech and build a more humanistic AI for the future.

Other industries have been reformed, and transformed, not through government regulation alone, but by empowered consumers. The movement toward organic, sustainably produced food is a phenomenon propelled forward by people who care. Consumers, especially Millennials, are willing to pay more for products that are sourced from fair trade, non-GMO ingredients that were not tested on animals and were manufactured in a humane and sustainable way. There are no laws compelling companies to do these things. Still, those companies that do will succeed in burnishing their brand in the eyes of consumers; those that don't will appear to be uncaring.

At Affectiva, we established a new field of AI for the purpose of humanizing technology and promoting its ethical development and deployment. We work diligently to make sure our technology is not created or implemented in a way that is harmful to society, perpetuates bias, or aggravates inequality. Why shouldn't every AI company follow these standards?

Surely, the public would rather use AI tools from sources that do not allow their technology to be used for widespread surveillance of populations, to spy on and harass people. In an era when racial and gender bias and economic inequality are at the forefront of the public conversation, I believe consumers, if given a clear choice, will opt for AI that does not perpetuate these behaviors, that is mindful of unconscious bias and collects data from diverse groups. Today, consumers have the ability to support "ethically sourced" and "sustainable" products over those of dubious provenance. We can opt for ethically sourced diamonds and chocolate, fair trade coffee, and even ethical fashion. So, why not add another industry to the list of companies that follow ethical and sustainable practices? Ethical technology.

One idea we are exploring is creating new criteria for "sustainable and ethical technology," a seal of approval issued by an independent group composed of technologists, ethicists, privacy advocates, and other interested parties for products that are developed or produced and deployed in an ethical way. When I shop for groceries, I go out of my way to buy organic, and I know a product is organic by the sticker on each piece of fruit or the package that provides me with a rating. I believe that organic is better not only for my family, but for the environment. In some cases, organic may not be as "perfect" as the absolutists would like, but it is my best choice.

Similarly, consumers should have the right to select AI-derived products that make a real commitment to meeting their ethical and moral standards. As an industry that is fast-moving and evolving so much, we have just begun to develop this concept, and there is much to be worked out. Admittedly, it is a complicated issue. At Affectiva, we are asking others in the tech community, and consumers and other stakeholders, to join us. AI takes many forms, with numerous and varied applications. We understand: There can't be one set of operating standards for everyone, but there can be basic principles, best practices, to which we all comply.

THE PATH FORWARD

How hard is it for a tech company to be honest and forthright? Affectiva has done it millions of times.

First and foremost, it is important that AI providers be transparent about how they collect data—what exactly they are collecting, how they are storing the data, and how that data will be used and by whom. This information should be available to people in simple, plain language that addresses those basic questions with transparency and clarity. It should not be buried in a sixty-page click-through license agreement written in legalese that the average consumer won't bother to read and wouldn't understand if they did.

We recognize that in a modern, connected world with "smart" objects everywhere, there are situations in which it isn't possible to get *explicit* consent, to ask people directly to opt in. For example, let's say you spot Pepper the social robot at a shopping mall and you walk over to engage it in conversation. Walking over to the robot is considered *implicit consent*: You are choosing to have this encounter. This is a very different scenario from one in which a retail mall hides cameras with Emotion AI in stores for the purpose of informing a salesperson what particular products capture your attention. That is intrusive and a violation of privacy, and I believe that the public is able to understand the distinction.

Second, the company must make its best effort to avoid algorithmic bias. Algorithmic bias is a reflection of a larger issue: Humanity itself is inherently biased. As a starting point, we need to strive for AI to be less biased than people are. With some effort, a company can eliminate bias, or at least significantly reduce it, by being mindful of how it acquires its data, how it trains and validates its models, and most of all, by building diversity into its teams.

We each have biases and blind spots based on our personal beliefs and experiences, and we each solve for the problems we know. Even with good intentions, people in a group developing algorithms fall into a similar demographic and come from similar backgrounds, and may unwittingly introduce bias. That's why companies need teams that are diverse in age, cultural background, ethnicity, life experiences, education, and other factors. Only when teams are diverse can we say, "You know, I noticed that there aren't enough data on people with my skin color. Can we make sure we include that?" Or "I have a beard, and I notice that we don't have any people in this data set who have beards." In our early days, Affectiva's data labeling team in Cairo actually flagged that we, at the time, did not have any data on women wearing a hijab, so we set out to add that to our data set. Diverse teams also have the potential to think of new applications for their technology that are representative of different groups, and to solve challenges for different groups of people.

Third, AI needs to be deployed in an ethical way: AI is not evil. The technology itself is neutral, but it may be used by people for nefarious purposes. We (software developers) have a responsibility to be highly selective about whom we allow to use our technology, and how we allow it to be used. This is where consumers can flex their muscles: Do you want to buy a product from a company that allows its AI to be used to spy on ethnic minorities, as it is used in China?

Consumers have more power than they think to control abuse in this sector. They just haven't had the tools to take action yet.

Big Tech companies are beginning to recognize that if they don't take action to prevent abuses, local and federal governments will. Certainly, some regulation is required. But these issues are so complex, and the technology moving so quickly, that the industry itself has to step in and take a proactive role in designing a strategy that doesn't inhibit progress, but also doesn't achieve it at the cost of privacy.

We started our annual Emotion AI Summit with the goal of building an ecosystem of ethicists, academics, and AI innovators and practitioners across industries to take action. The first year, our theme was "human connection"; the following year we explored "trust in AI." In 2019, the theme is "human-centric AI": How can we ensure that AI is designed, developed, and deployed with the end user in mind? Throughout the day at these summits, we interweave conversations on the ethics of AI and always close the day with a panel on ethics.

As a society, we are just starting to have a conversation about the role of AI in our lives, and how to use it ethically and fairly for the betterment of humanity. We can't let the presence of some bad players exploiting this technology thwart the ability of the good players to create tools and services that are helpful to society. We have to establish standards, and those who cross the line, either directly or indirectly (by licensing technologies to bad players), need to know that the tables have been turned: Consumers are now tracking them and will penalize companies that don't conform to basic humane standards.

I've devoted my career to bringing emotional intelligence to computers to create technology that is responsive to the needs of human beings. My company is a reflection of these values. Since its inception more than a decade ago, we have made a strong commitment to building technology in an ethical and moral way that respects the rights of individuals. We carefully choose whom we work with, and we withhold our technology from people and companies who don't adhere to our standards. As long as we are in business, we will continue to do so, and we will push the industry in that same direction.

After all, shouldn't we put the *human* before artificial?

Afterword

The quest to create more human centric technology and a more empathetic world has brought me to where I am now.

In 2019, we celebrated Affectiva's tenth anniversary where it all began, at the MIT Media Lab. It was a family affair. We invited a diverse group of friends and supporters of the company, from our earliest investors, advisers, and partners to our most recent ones. These people believed in us—in me—even when I didn't believe in myself. We had the first Affectiva employees and the current Affectiva team together with their spouses, partners, and children of all ages. A few strollers (and toddlers) were roaming around. Jana and Adam were there and so was my co-founder, Rosalind Picard. As I gave my keynote, I realized that of all our accomplishments, I am most proud of the professional and personal growth of each person in the room. Through the years, members of our team have gotten married and become parents, and some, like me, had to adjust to life in a new country and a few became new Americans. Some were faced with the challenges of caring for older parents; some experienced the loss of loved ones. We shared one another's triumphs and traumas, as we've moved the field of Emotion AI forward together. I feel blessed and grateful to be part of their journey.

Twenty years ago, when I broke ground on this new field, it was awkward to talk about emotions. It made my peers uncomfortable. Now the world has changed; people see the importance of emotions in health and well-being, in how we make decisions, and in how we communicate. We no longer have to find acceptable synonyms for *emotion;* we now use the "e-word."

When I was younger (a nice Egyptian girl) I hid my emotions from others, and for that matter, often didn't even acknowledge my own emotions. I was emotionally dependent, stifled by fear of "What will the neighbors think!" But now I am emotionally (and financially) independent. I've discovered that the more I can embrace my emotions, be honest about them to myself, and share them with others, the more that invites people to reciprocate and share.

At Affectiva's 2019 Emotion AI Summit, my mom was visiting from the UAE. I invited her up onstage with me and introduced her to the crowd as my greatest supporter and mentor. I felt a rush of love and gratitude toward this woman who had given so much of herself to my sisters and me, and I can honestly say, without her, I wouldn't be where I am today. I teared up in public. The younger me would never have done that; I would never have opened up that way and made myself so vulnerable, nor, as I have done in this book, would I have ever talked publicly about my family relationships, the dark times when I was going through my divorce, or my struggles to become CEO. But I have learned that being attuned to our emotions, and being unafraid to show and act on them, is empowering. That's how you build strong personal connections versus putting up a barrier, whether in the digital or real world. It creates true empathy; we become more understanding and accepting of ourselves and of others.

As much as I have gained insight into my own emotional life, I am still a work in progress. At times, I still have difficulty reconciling the nice Egyptian girl with the American entrepreneur who's emotionally independent. I've shed my headscarf. I look "American" in the way that I speak and dress, yet my brain is still "Egyptian" because part of me can't believe that an Egyptian woman can do all of this. I still hear the "Debbie Downer" naysayer voice in my head, but I don't allow it to hold me back. I have learned to reframe the message; it is now my advocate, not my adversary, challenging me to move forward out of my comfort zone. It energizes me to work harder to exceed my expectations and the expectations of others.

Twenty years from now, I am confident that we will be interacting with our technology on our own human terms: We will be communicating with one another using the full spectrum of emotion. Every digital interaction, whether it's a text, a tweet, a voice, a video message—or whatever comes next—will have Emotion AI built in. So, for example, imagine if we're able to quantify emotion and have an emotion metric: Imagine receiving a tally of your emotional impact at the end of the day—"150 people empathized with you today!" Or imagine receiving an alert if a carelessly written text offended someone; even better, imagine if your virtual assistant intervenes before you send out an offending message. "That's really nasty, do you want to send this?" Recognizing emotion in the cyberworld will keep emotion front and center in our lives.

No one can predict exactly how technology will unfold; there are always some surprises. But when I think of what's to come for my kids and for Affectiva, I envision a world that is full of compassion and empathy, where people from all corners of the planet can not only communicate with one another with and through technology but do so in a way that doesn't erode our humanity—but rather makes us more caring, better human beings.

Acknowledgments

I am a big believer that it takes a village to get anything done! This book is no different. I am so grateful for my village without whom this book would not have been possible.

First, I want to thank my coauthor, Carol Colman. I first met Carol while I was a postdoc at the MIT Media Lab. We clicked right away! Years later, we ran into each other at a health conference. Carol suggested I write a book. I scoffed, adding, "I don't have anything to say—what would I possibly write about!" Thank you, Carol, for seeing my story and believing in it. Thank you for embarking on this journey with me, for holding me accountable to a timeline, for working tirelessly to make sure everything was perfect, and for challenging me to dig deep and be vulnerable. It still amazes me how Carol has gotten to know my voice so well, so much so that we would spook our editor and publishing team. I am grateful for this wonderful partnership.

I am also grateful to my "book crew." Laurie Bernstein, my agent, for relentlessly advocating on my behalf and for patiently teaching me Book Publishing 101! Roger Scholl for being an amazing editor—I still remember the first time we met over lunch and he started with a caveat: "I do not take on AI books anymore." Roger then proceeded to ask about my personal story and had an aha moment: that the book should target a wider audience and tell my journey alongside the creation of Emotion AI. Thanks to Erin Little for the thoughtful edits and for appreciating my quirky sense of humor! Tina Constable for having big ambitions for this book; despite having published many very well-known people over thirty years, Tina still gets excited about publishing "emerging" voices like mine. And the rest of the Currency publishing team, including Campbell Wharton, the associate publisher; Cindy Murray, the head of publicity; Melanie DeNardo, who handled the day-to-day publicity for the book within the Random House Group; Andrea DeWerd, the head of marketing; the sales team at the Random House Group; and the incredible team at the Random House Group who all read the book and encouraged us to "pivot" to maximize reach—Susan Corcoran, Todd Berman, and Leigh Marchant—and who continue to support *Girl Decoded* with passion and enthusiasm. Mark Fortier and Lauren Kuhn for indulging me on my ambitious plans for getting the book out. I am forever grateful.

As this book journey unfolded, it became clear that there were many moving parts, too many for me to handle while running Affectiva. I needed an A team to take this on! And what an A team! Hailey Melamut: Thank you for the incredible partnership we have had over the years; you somehow have cracked my voice and dare I say my brain, too. Hailey took on the role of book launch manager, and I couldn't have picked a better person: organized, strategic, and attentive to all the details. Thanks to the full March Communications team for partnering to get our story out into the world. Iulia Nandrea-Miller, my executive assistant and my chief yes officer: Iulia, you are a true blessing—being the EA of an energizer bunny who has a high bar for quality is not easy. Thank you for organizing my life, taking on all aspects related to the book, and for just getting stuff done. It continues to be a pleasure to work and grow together. Rula el Kaliouby, my brand and social media manager and strategist—and my baby sister. What a genius you are. I love how you've nailed the essence of what this book is about and who I am, and how to engage the world—young and old, men and women, from all corners of the universe. Gabi Zijderveld, Affectiva's chief marketing officer, friend, mentor, and my partner in crime. Oh and my chief no officer! Gabi, who is Dutch American, singlehandedly brings the stereotype of the Dutch being honest to reality! Gabi will say it as it is and that's why I love her. She's built Affectiva and my personal brand; she and I created and seeded the Emotion AI category and continue to paint this vision of an emotion-enabled world. Gabi read multiple versions of this book and has kept me honest throughout this process.

I have been so fortunate to have met people who supported and mentored me throughout my career. My cofounder, mentor, and role model Rosalind Picard took a risk on a young, Egyptian Muslim woman who had to commute from Cairo as a postdoc. Thank you, Roz! For cofounding Affectiva with me. For defining our North Star and our core values from the onset. For teaching me never to take no for an answer. For showing me that you can be a kickass professional but also be an amazing wife, mother, mentor, and more. For showing me that faith transcends all religions and disciplines. Peter Robinson, for pushing me in my research outside of my comfort zone and for being my home away from home. Simon Baron-Cohen for providing access to his data and his team. My mentors and faculty at my beloved alma mater, the American University in Cairo, for being the bridge to the Western world and showing me that the world has no limits.

I have also been blessed with meeting amazing people who have shaped and challenged the ideas in this book. I am grateful to the domain experts who agreed to be interviewed for this book: Ola Bostrom, Kate Darling, Joe Dusseldorp, Ehsan Hoque, Karl Iagnemma, Cory Kidd, Loren Larsen, Tim Leberecht, Taniya Mishra, Bryan Reimer, Ned Sahin, Erin Smith, Steven Vannoy, Ben Waber, Peter Weinstock, and Matt Willis. Cynthia Breazeal for being an amazing role model. Adam Grant, Erik Brynjolfsson, and Richard Yonck for early feedback on the book. Lots of gratitude to the journalists who have written about my work and made it accessible to the general public, who, from the early days, have held us accountable and urged us to think through the implications of our technology and be vocal advocates for the ethical development and deployment of AI. Thank you to Raffi Khatchadourian, whose 2015 article "We Know How You Feel" in *The New Yorker* put us on the map. June Cohen for putting me on the TED stage. Dale DeLetis for coaching me on how to speak in public. Debbie Simon for showing me, and my kids, the power of having a voice and speaking your truth. The Stern team, especially Danny Stern, Mel Blake, and Ania Trzepizur, for the opportunity to share my work across the world.

This book is not just about hopes and dreams. Many of the concepts in this book have been made true by my team at Affectiva. Working alongside this team day in, day out, is a gift—the hard work, the passion for our mission of humanizing technology, the teamwork and collaboration, and the commitment to the ethical development and deployment of AI. There are a few people in particular without whom Affectiva and I would not be where we are today: David Berman who showed me what a grown-up company looks like even when there was only a handful of us. Nick Langeveld for giving me an opportunity to take the helm. Tim Peacock for being my trusted right hand. I first interviewed Tim in 2011, luring him to join Affectiva over Chinese lunch! That Chinese lunch must have been great because he joined as our VP Engineering and we haven't looked back. When I became CEO, Tim stepped in as my COO; Tim keeps me honest and is my go-to whenever I am at a crossroad. Gabi Zjiderveld for creating Affectiva's brand, doing all that by being scrappy but not crappy! Andy Zeilman for leading our strategy and stepping in wherever needed. Graham Page for being our very first executive champion and customer. Jay Turcot for the wonderful deep discussions on strategy, leadership, and how to achieve some semblance of work-life balance. Watching Jay grow personally and professionally continues to be the highlight of my career. Abdelrahman Mahmoud for teaching me when to become invested in someone's success—I will forever be your biggest advocate! Taniya Mishra for creating and building our flagship internship program, giving young people an opportunity to shape this brave new world of AI. Dan McDuff for our shared curiosity and for geeking out on the universality of human behavior and emotion expression.

As a venture-backed start-up, your investors are everything. I am super grateful to Hans Lindroth and Jeff Krentz, our early investors, for sticking with us as we've evolved, pivoted, and evolved again. Our current investors for believing in our mission: Wael Mohamed and Keith Foster for being my sounding board. Ollen Douglas and Maggie Dorn at Motley Fool Ventures for valuing integrity, diversity, and ethics. The Aptiv team, especially Kevin Clark, David Paja, Sean Valentine, Bob Bibby, and Andreas Heim, for selecting us to be partners for this grand vision of bringing safety to our roads. Our collaborators, partners, and clients who took this technology, sometimes in its very early days, and deployed it across the world in ways that are changing how we connect with technology and with one another, often in ways I hadn't ever imagined. Many banked their careers by championing our technology and tied their own professional success to ours. Thank you.

A start-up is like an emotional roller coaster—and sometimes it felt like I was riding it alone! I am grateful to my mentors and advisers who cheered me on when I needed it the most. Andy Palmer for being there from Day 1 and for raising the bar on what it means to support a founder, especially a woman founder. Frank Moss for planting the seed of starting a company. Ossama Hassanein, aka Dr. O, for showing me that my personal well-being matters, too. Eric Schurenberg for giving my voice a platform. Babak Hodjat, Karl Iagnemma, Danny Lange, and Bryan Reimer for being my thought partners and our advocates. Elia Stupka for challenging me to verbalize my true emotions. Max Tega Mark for helping me think big and not conform to convention. Eric Horvitz for the mentorship, as I went from academic to entrepreneur. Gregory Wilson for nudging me to step outside my comfort zone and for helping me visualize myself as CEO and thought leader.

I have also been fortunate enough to play the role of mentor to a number of people who, throughout that process, have taught me important lessons in life and love: Sara Bargal, Marwa Mahmoud, and Radwa Hamed for showing the world that Egyptian women are total kickass! Erin Smith for teaching me that while age doesn't matter, passion and perseverance do. It is a joy and a gift to play a small part in your amazing journeys.

And finally, my family. Where do I even begin? Writing this book was a process of intense self-reflection, reflecting on my upbringing and reconciling my culture and traditions with who I am today: an Egyptian American empowered scientist and entrepreneur. I owe every bit of where I am today to my family.

Thank you to my dad, Ayman el Kaliouby, for instilling in me the core values of hard work, compassion, and helping others. For investing in my education and for believing that his three daughters have great potential, even when people around him pitied him for not having any sons!

Thank you to my mom, Randa Sabry, for her unconditional love and unwavering support. There is nothing my mom hasn't done for me and my sisters and our kids. She took Jana for a full month while I finished my PhD dissertation. She will show up in Boston to help if I have intense traveling. My mom, one of the first programmers in the Middle East, is a role model of what it means to be a lifelong learner: At sixty plus, she still attends courses to learn new programming languages or new educational pedagogies, or she'll sign up for a new dance class. She shows us the joy of learning and teaches us how to embrace our emotions and how that can be magical.

Thank you to my sisters, Rasha and Rula. Rasha for being the glue of this family, for reminding me every day that when compassion comes first, the rest all works out. For always being there when the going gets tough. Rula for defying the status quo and doing so in style and with confidence! I've always been in awe of my youngest sister—while I was not allowed to date and had strict curfews, somehow Rula got away with it. I am grateful to my two nieces—Amina for reminding me so much of my younger self, and Zeina, at eight, for being the creative rebel I always wanted to be.

I am grateful to Wael, my first love and father of my kids, for putting up with a super-ambitious partner! My mother-in-law, Tant Laila, for leading with kindness, and my late Uncle Ahmed for loving me like his own daughter. My brother-in-law, Houssam, and sister-in-law, Sahar, for continuing to make me feel part of this family, even post-divorce.

Thank you to my kids, Jana and Adam, who are my North Stars, my thought partners, my travel companions, my best friends, my mentors, and my confidants. Jana for inspiring me every single day and for showing me how to be an empowered young woman with a powerful voice. Jana, I can't wait to see what you will do in the world. Adam for showing me how to get up and learn from my failures, for reminding me to take a deep breath and take short breaks, for making me laugh, and for proving that kindness always wins. Jana and Adam, you not only bring me pride and tremendous joy but you keep me grounded and hold me accountable. I am blessed to learn and grow alongside you two. I hope that this book inspires you to do good in the world. May we always be this close, and may we continue to find joy in this incredible and exciting journey together, without attaching to outcomes.

Bibliography

The entries in this bibliography are listed in the order in which they are used in the book.

Introduction: Emotion Blind

Ellis, Ralph, Nick Valencia, and Devon Sayers. "Chief to Recommend Charges Against Florida Teens Who Recorded Drowning." CNN, July 22, 2017. https://edition.cnn.com/2017/07/21/us/florida-teens-drowning-man/index.html.

"No Charges for 5 Teens Who Mocked Drowning Man, Didn't Help." AP News. Information from *Florida Today* (Melbourne, FL), June 23, 2018.

Goleman, Daniel. *Emotional Intelligence: Why It Can Matter More Than IQ*. New York: Bantam Books, 1995.

"About Three in Ten U.S. Adults Say They Are 'Almost Constantly' Online." Pew Research Center, July 25, 2019. https://www.pewresearch.org/fact-tank/2019/07/25/americans-going-online-almost-constantly/.

"There Will Be 24 Billion IoT Devices Installed on Earth by 2020." Business Insider Intelligence, June 9, 2016. https://www.businessinsider.com/there-will-be-34-billion-iot-devices-installed-on-earth-by-2020-2016-5.

Chapter 5: The Spark

Rosalind Picard. *Affective Computing*. Cambridge, MA: MIT Press, 1997.

Antonio Damasio. *Descartes' Error: Emotion, Reason, and the Human Brain*. New York: G. P. Putnam's Sons, 1994.

Ekman, Paul, and Wallace V. Friesen. *Facial Action Coding System: A Technique for the Measurement of Facial Movement*. Palo Alto, CA: Consulting Psychologists Press, 1978.

"Increase Your Emotional Awareness and Detect Deception." Online training course. https://www.paulekman.com.

Kanade, Takeo, Jeffrey F. Cohn, and Yingli Tian. "Comprehensive Database for Facial Expression Analysis." *Proceedings of the Fourth IEEE International Conference on Automatic Face and Gesture Recognition (FG '00)*, Grenoble, France, March 2000, pp. 46–53. http://www.cs.cmu.edu/~face/Papers/database.PDF.

Chapter 6: A Married Woman

Rana el Kaliouby. "Enhanced Facial Feature Tracking of Spontaneous Facial Expression." M.Sc. diss., American University in Cairo, 2000.

Amr Khalid, "The Hijab." Translated and transcribed from a lecture. http://www.oocities.org/mutmainaa5/articles/hijab1.html.

Chapter 7: Stranger in a Strange Land

"History." Newnham College, University of Cambridge. https://www.newn.cam.ac.uk/about/history/.

Chapter 8: A Mad Scientist Talks to a Wall

Baron-Cohen, Simon, Sally Wheelwright, et al. "The 'Reading the Mind in the Eyes' Test, Revised Version: A Study with Normal Adults and Adults with Asperger Syndrome or High-Functioning Autism." *Journal of Child Psychology and Psychiatry and Allied Disciplines* 42, no. 2 (2001): 241–51.

Baron-Cohen, Simon. "Reading the Mind in the Eyes Test." https://www.autismresearchcentre.com/arc_tests/.

Centers for Disease Control and Prevention (CDC). "Data and Statistics on Autism Spectrum Disorder." CDC. https://www.cdc.gov/ncbddd/autism/data.html.

Baron-Cohen, Simon. *Mind Reading: The Interactive Guide to Emotions*. London: Jessica Kingsley Publishers, 2003.

Chapter 9: The Challenge

Baron-Cohen, Simon. "Autism and the Technical Mind." *Scientific American*. November 2012, pp. 307 and 72–75.

———. "The Essential Difference: The Male and Female Brain." *Phi Kappa Phi Forum* 45, no. 1 (January 2005).

———. "The Hyper-systemizing, Assortative Mating Theory of Autism." *Progress in Neuro-Psychopharmacology and Biological Psychiatry* 30 (2006): 865–72.

Baron-Cohen, Simon, and Sally Wheelwright. "The Empathy Quotient (EQ): An Investigation of Adults with Asperger Syndrome or High Functioning Autism, and Normal Sex Differences." *Journal of Autism and Developmental Disorders* 34 (2004): 163–75.

Baron-Cohen, Simon, Sally Wheelwright, et al. *The Exact Mind: Empathising and Systemising in Autism Spectrum Conditions*. Oxford: Blackwell, 2002.

Chapter 10: Learning Human

el Kaliouby, Rana, and Peter Robinson. "Real-time Inference of Complex Mental States from Facial Expressions and Head Gestures." Paper presented at the IEEE International Workshop on Real-time Computer Vision for Human–Computer Interaction at CVPR, January 2004.

Chapter 11: Mommy Brain

el Kaliouby, Rana, and Peter Robinson. "FAIM: Integrating Automated Facial Affect Analysis in Instant Messaging." In Nuno Jardim Nunes and Charles Rich, eds. *Proceedings of the International Conference on Intelligent User Interfaces 2004,* Funchal, Madeira, Portugal, January 13–16, 2004, pp. 244–46. http://doi.acm.org/10.1145/964442.964493.

el Kaliouby, Rana. "Mind-Reading Machines: Automated Inference of Complex Mental States." PhD diss., University of Cambridge Computer Laboratory, 2005.

Chapter 12: Crazy Ideas

el Kaliouby, Rana, Rosalind Picard, and Simon Baron-Cohen. "Affective Computing and Autism." *Annals of the New York Academy of Sciences* 1093 (2006): 228–48. doi:10.1196/annals.1382.016.

el Kaliouby, Rana, Alea Teeters, and Rosalind W. Picard. "An Exploratory Social-Emotional Prosthetic for Autism Spectrum Disorders." International Workshop on Wearable and Implantable Body Sensor Networks (BSN '06), Cambridge, MA, April 3–5, 2006, pp. 2 and 4. doi: 10.1109/BSN.2006.34.

Jennifer Schuessler. "The Social-Cue Reader." *New York Times Magazine*. December 10, 2006.

Negroponte, Nicholas. *Being Digital*. New York: Alfred A. Knopf, 1999.

Teeters, Alea. "Use of a Wearable Camera System in Conversation: Toward a Companion Tool for Social-Emotional Learning in Autism," MS thesis, MIT, September 2007.

Ahn, H. I., A. Teeters, A. Wang, C. Breazeal, and R. W. Picard. "Stoop to Conquer: Posture and Affect Interact to Influence Computer Users' Persistence," The Second International Conference on Affective Computing and Intelligent Interaction, Lisbon, Portugal, September 12–14, 2007.

Ahn, H. I., and R. W. Picard, "Measuring Affective-Cognitive Experience and Predicting Market Success." *IEEE Transactions on Affective Computing*, June 2014.

Chapter 14: Demo or Die

Frank Moss. *The Sorcerers and Their Apprentices.* New York: Crown Business, 2011.

Madsen, Miriam, Rana el Kaliouby, et al. "Technology for Just-in-Time In Situ Learning of Facial Affect for Persons Diagnosed with an Autism Spectrum Disorder." *Proceedings of the Tenth ACM Conference on Computers and Accessibility (ASSETS)*, Halifax, Nova Scotia, October 13–15, 2008.

Teeters, Alea, Rana el Kaliouby, et al. "Novel Wearable Apparatus for Quantifying and Reliably Measuring Social-Emotional Expression Recognition in Natural Face-to-Face Interaction." Poster at International Meeting for Autism Research (IMFAR), London, May 15–17, 2008.

Madsen, Miriam, Rana el Kaliouby, et al. "Lessons from Participatory Design with Adolescents on the Autism Spectrum." Conference on Human Factors in Computing Systems (CHI '09), Boston, MA, April 4–9, 2009.

Chapter 16: My Arab Spring

"Egypt Erupts in Jubilation as Mubarak Steps Down." *New York Times*, February 11, 2011.

McDuff, Dan J., Rana el Kaliouby, and Rosalind W. Picard. "Crowdsourcing Facial Responses to Online Videos." *IEEE Transactions on Affective Computing* 3 no. 4 (2012): 456–68.

"Onslaught." Dove commercial, 2008. https://www.youtube.com/watch?v =9zKfF40jeCA.

"Geyser." Huggies Disposable Diapers commercial, 2008. https://www.youtube .com/watch?v=AVRpE7982Js.

"Joy Is BMW." BMW commercial, 2009. https://youtu.be/oR4wkZM9Zis.

McDuff, Dan J., Rana el Kaliouby, et al., "Predicting Ad Liking and Purchase Intent: Large-scale Analysis of Facial Responses to Ads." *IEEE Transactions on Affective Computing*, July 2015. https://affect.media.mit.edu/ pdfs/14.McDuff_et_al-Predicting.pdf.

McDuff, Dan J., Rana el Kaliouby, et al. "Predicting Online Media Effectiveness Based on Smile Responses Gathered Over the Internet." Tenth IEEE International Conference on Automatic Face and Gesture Recognition, Shanghai, China, April, 2013.

McDuff, Dan J. "Crowdsourcing Affective Responses for Predicting Media Effectiveness." PhD thesis, MIT, June 2014.

McDuff, Dan J., J. M. Girard, and Rana el Kaliouby. "Large-scale Observational Evidence of Cross-cultural Differences in Facial Behavior." *Journal of Nonverbal Behavior* 41, no. 1 (2017): 1–19.

Chapter 18: Woman in Charge

Adler, Jerry. "Smile, Frown, Grimace, and Grin—Your Facial Expression Is the Next Frontier in Big Data." *Smithsonian Magazine,* December 2015.

Smith, Aaron. "Nearly Half of American Adults Are Smartphone Owners." Pew Research Center, Internet and Technology, 2012. https://www.pewinternet.org/2012/03/01/nearly-half-of-american-adults-are-smartphone-owners/.

Chapter 19: Hacking the Hackathon

Subbaraman, Nidhi. "Affectiva Invites Local Developers to Test Emotion-Sensing Tech: Hackathon Fills Up with Mixed-Gender Group." *Boston Globe,* February 1, 2016.

Curtin, Sally C., and Melonie Heron. "Death Rates Due to Suicide and Homicide Among Persons Aged 10–24: United States, 2000–2017." *NCHS Data Brief,* No. 352, October 2019. https://www.cdc.gov/nchs/data/databriefs/db352-h.pdf.

National Institute of Mental Health (NIMH). "Suicide." Mental Health Information: Statistics, NIMH, n.d. https://www.nimh.nih.gov/health/statistics/suicide.shtml.

World Health Organization (WHO). *The World Health Report 2001—Mental Health: New Understanding, New Hope.* NMH Communications, WHO, October 2001. https://www.who.int/whr/2001/en/.

Khatchadourian, Raffi. "We Know How You Feel: Computers Are Learning to Read Emotion, and the Business World Can't Wait." *The New Yorker,* January 12, 2015.

Vannoy S., S. Gable, M. Brodt, et al. "Using Affect Response to Dangerous Stimuli to Classify Suicide Risk." Paper presented at CHI 2016 Computing and Mental Health Workshop, San Jose, CA, May 8, 2016. http://alumni.media.mit.edu/~djmcduff/assets/publications/Vannoy_2016_Using.pdf.

American Psychiatric Association. *Diagnostic and Statistical Manual of Mental Disorders (DSM-5).* Washington, DC: American Psychiatric Association, 2013.

Chapter 20: Gone Quiet

Vahabzadeh, A., N. Y. Keshav, J. P. Salisbury, and N. T. Sahin. "Improvement of Attention-Deficit/Hyperactivity Disorder Symptoms in School-aged Children, Adolescents, and Young Adults with Autism via a Digital Smart-glasses-Based Socioemotional Coaching Aid: Short-Term, Uncontrolled Pilot Study." *JMIR Mental Health* 5, no. 2 (March 2018): e25. doi: 10.2196/mental.9631.

Chapter 21: Secrets of a Smile

Dusseldorp, J. R., D. L. Guarin, M. M. van Veen, et al. "In the Eye of the Beholder: Changes in Perceived Emotion Expression After Reanimation." *Plastic and Reconstructive Surgery* 144, no. 2 (August 2019): 457–71.

Smith, Erin. "Forbes 30 Under 30." https://www.forbes.com/profile/erin -smith/.

Mhyre, T. R., J. T. Boyd, R. W. Hamill, and K. A. Maguire-Zeiss. "Parkinson's Disease." *Subcellular Biochemistry* 65 (2012): 389–455.

Chapter 23: Leveling the Playing Field

Buranyi, Stephen. "How to Persuade a Robot That You Should Get the Job." *The Guardian,* March 4, 2018.

Gerdeman, Dina. "Minorities Who 'Whiten' Job Résumés Get More Interviews." Harvard Business School (website), May 17, 2017. https:// hbswk.hbs.edu/item/minorities-who-whiten-job-resumes-get-more -interviews.

Turban, Stephen, Laura Freeman, and Ben Waber. "A Study Used Sensors to Show That Men and Women Are Treated Differently at Work." *Harvard Business Review,* October 23, 2017.

Fung, Michelle, Yina Jin, RuJie Zhao, and Mohammed (Ehsan) Hoque. "ROC Speak: Semi-Automated Personalized Feedback on Nonverbal Behavior from Recorded Videos." *Proceedings of the 2015 ACM International Joint Conference on Pervasive and Ubiquitous Computing (UbiComp '15).* New York: ACM, 2015, pp. 1167–78. https://doi.org/10.1145/2750858 .2804265.

Samrose, Samiha, Ru Zhao, Jeffery White, Vivian Li, Luis Nova, Yichen Lu, Mohammad Ali, and Ehsan Hoque. "CoCo: Collaboration Coach for Understanding Team Dynamics During Video Conferencing." *Proceedings of the ACM on Interactive, Mobile, Wearable, and Ubiquitous Technologies* 1, no. 3 (2018): 1–24. https://dl.acm.org/citation.cfm?doid=3178157 .3161186.

Chapter 24: Human-*ish*

Time Staff. "The 25 Best Inventions of 2017: A Robot You Can Relate To." *Time,* December 1, 2017.

Van Camp, Jeffrey. "My Jibo Is Dying and It's Breaking My Heart." *Wired,* March 8, 2019.

Jonze, Spike (director). *Her.* Warner Bros. Pictures, 2013. https://www.imdb .com/title/tt1798709/.

Darling, Kate. "Why We Have an Emotional Connection to Robots." TED, October 2018. https://www.ted.com/speakers/kate_darling.

Ackerman, Evan. "Kids Love MIT's Latest Squishable Social Robot (Mostly): Tega Uses Cuteness and Artificial Intelligence to Teach Spanish to Preschoolers." *IEEE Spectrum,* March 2, 2016.

Petersen, Sandra, Susan Houston, Huanying Qin, Corey Tague, and Jill Studley. "The Utilization of Robotic Pets in Dementia Care." *Journal of Alzheimer's Disease* 55, no. 2 (2017): 569–74.

Johnson, Khari. "Pfizer Launches Pilot with Home Robot Mabu to Study Patient Response to AI," *VentureBeat,* September 12, 2019.

Lucas, Gail M., Albert Rizzo, Jonathan Gratch, et al. "Reporting Mental Health Symptoms: Breaking Down Barriers to Care with Virtual Human Interviewer." *Frontiers in Robotics and AI* 4, no. 51 (2017).

Neiman, Andrea B., Todd Ruppar, Michael Ho, et al. "CDC Grand Rounds: Improving Medication Adherence for Chronic Disease Management: Innovations and Opportunities." *Morbidity and Mortality Weekly Report* 66, no. 45 (November 17, 2017).

"Meet Ellie: The Robot Therapist Treating Soldiers with PTSD." USC Institute for Creative Technologies, October 1, 2016. http://ict.usc.edu/news/ meet-ellie-the-robot-therapist-treating-soldiers-with-ptsd/.

Kleinsinger, Fred, MD. "The Unmet Challenge of Medication Nonadherence." *The Permanente Journal* 22 (2018): 18–33 (July 5, 2018).

Johnson, Khari. "Softbank Robotics Enhances Pepper the Robot's Emotional Intelligence," *VentureBeat,* August 28, 2018.

Chapter 25: Alexa, We Need to Talk

"The Smart Home Is Creating Frustrated Consumers: More than 1 in 3 US Adults Experience Issues Setting Up or Operating a Connected Device." *Business Wire,* January 30, 2018.

Mishra, Taniya. "Decomposition of Fundamental Frequency Contours in the General Superpositional Intonation Model." Diss., Oregon Health and Science University, Department of Science and Engineering, 2008.

Maor, Elad, D. Sara Jaskanwal, Diana M. Orbelo, et al. "Voice Signal Characteristics Are Independently Associated with Coronary Artery Disease." *Mayo Clinic Proceedings* 93, no. 7 (July 2018): 840–47. https://doi.org/10.1016/j.mayocp.2017.12.025.

Hakkani-Kur, Dilak. "Alexa Prime Social Challenge—Grand Challenge 3." *Alexa Blogs,* March 4, 2019. https://developer.amazon.com/blogs/alexa/post/c025d261-e14f-403d-ba5d-b20f8fc86914/alexa-prize-socialbot-grand-challenge-3-application-period-now-open.

Chapter 26: Robots on Wheels

National Highway Traffic Safety Administration (NHTSA). "Automated Vehicles for Safety." NHTSA. https://www.nhtsa.gov/technology-innovation/automated-vehicles-safety.

Calvert, Scott. "Pedestrian Deaths Reach Highest Level in Nearly 30 Years." *Wall Street Journal,* February 28, 2019.

National Highway Traffic and Safety Administration (NHTSA). "Distracted Driving." NHTSA. https://www.nhtsa.gov/risky-driving/distracted-driving.

Centers for Disease Control and Prevention (CDC). "Distracted Driving." CDC. https://www.cdc.gov/motorvehiclesafety/distracted_driving/index.html.

Lienart, Paul. "Most Americans Wary of Self-Driving Cars: Reuters/Ipsos Poll." Reuters, January 29, 2018.

Chapter 27: Human Before Artificial

Buckley, Chris, Paul Mazur, and Austin Ramzy. "How China Turned a City into a Prison." *New York Times,* April 4, 2019.

Nielsen. "Was 2018 the Year of the Influential Sustainable Consumer?" N (website). December 17, 2018. https://www.nielsen.com/us/en/insights/article/2018/was-2018-the-year-of-the-influential-sustainable-consumer/.

Index

PENGUIN PARTNERSHIPS

Penguin Partnerships is the Creative Sales and Promotions team at Penguin Random House. We have a long history of working with clients on a wide variety of briefs, specializing in brand promotions, bespoke publishing and retail exclusives, plus corporate, entertainment and media partnerships.

We can respond quickly to briefs and specialize in repurposing books and content for sales promotions, for use as incentives and retail exclusives as well as creating content for new books in collaboration with our partners as part of branded book relationships.

Equally if you'd simply like to buy a bulk quantity of one of our existing books at a special discount, we can help with that too. Our books can make excellent corporate or employee gifts.

Special editions, including personalized covers, excerpts of existing books or books with corporate logos can be created in large quantities for special needs.

We can work within your budget to deliver whatever you want, however you want it.

**For more information, please contact
salesenquiries@penguinrandomhouse.co.uk**